Fluid Power
Educational
Series

Hydraulic Power Packs
(In the SI Units)

Joji Parambath

Hydraulic Power Packs
(In the SI Units)

Copyright © 2020 Joji Parambath

All rights reserved

ISBN: 9798653661006

https://jojibooks.com

Disclaimer of Liability

The contents of this book have been checked for accuracy. Since deviations cannot be precluded entirely, we cannot guarantee full agreement. Only qualified personnel should be allowed to install and work on hydraulic equipment. Qualified persons are defined as persons who are authorised to commission, to ground, and tag circuits, equipment, and systems following established safety practices and standards.

Dedicated to

all my relatives and friends in Thalasseri

Table of Contents

PREFACE

As the usage of hydraulic systems is becoming more widespread, there is a greater need for understanding the function and operation of hydraulic power packs. This book takes up a detailed discussion of hydraulic power packs and their constituent parts including reservoirs, pumps and pressure relief valves. This book also gives a brief note on the topic of heat dissipation and sound reduction techniques in hydraulic systems. The book uses the SI system of units.

Many other fluid power topics are given in other textbooks under the fluid power educational series by the same author. A list of all the books is given at the end of the book (Page No. 79). Also, please see the details at https://jojibooks.com

Enjoy reading the book.
Your feedback is most welcome.

JOJI Parambath

Chapter 1 | Hydraulic Power Packs -Introduction

A hydraulic system requires a sufficient amount of high-quality fluid at all times for its efficient operation. A power pack, as shown in Figure 1.1, is the unit that supplies the required fluid to all actuators in the system. It is a compact, portable, and custom-designed or pre-engineered assembly consisting of essential and optional components. The essential components are a fluid-filled reservoir, close-coupled pump-motor unit, pressure relief valve, and pressure gauge, and the optional components include a heat exchanger, temperature controller, directional control valves, filters, etc. It also consists of necessary instrumentation, and other accessories, such as accumulators, hoses, and quick-disconnect couplings. The modern way of configuring a power pack is from standardised sub-assemblies.

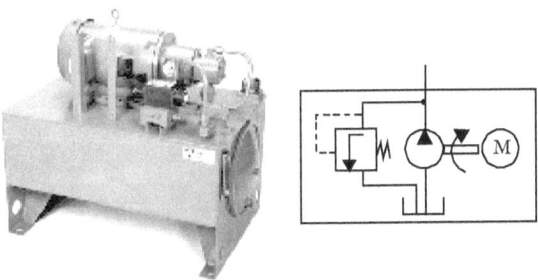

Figure 1.1 | Hydraulic power pack
Courtesy: Advance Motion Control, USA

The power packs are designed to operate with mineral or fire-resistant or bio-degradable hydraulic fluids of correct viscosity grade and viscosity index. The choice of a power pack is decided by many factors like the pressure, flow rate, the availability of space and the amount of heat involved in the application. Another factor to be reviewed is whether the unit is to be used in a factory or marine environment, or hazardous location.

Chapter 2 | Hydraulic Reservoirs

A reservoir (tank) stores and conditions a given quantity of hydraulic fluid. A well-designed reservoir:

- allows a reasonable dwell time for the fluid
- allows most of the contaminants to settle
- assists in dissipating the heat
- allows air bubbles to dissipate
- compensates for the fluid volume changes
- provides a convenient mounting place for the pump-motor unit, valves, and other components

A reservoir is designed, sized, and constructed as per the requirements of the application. It is equipped with many accessories to meet the special requirements of the application.

Constructional Features

Figure 2.1 shows the cross-sectional view of a reservoir.

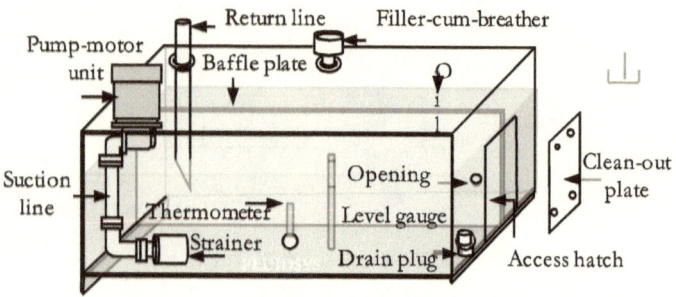

Figure 2.1 | A cross-sectional view of a hydraulic reservoir

A well-designed hydraulic reservoir should be completely enclosed and self-contained. It should have sufficient capacity to supply the required quantity of fluid to the associated system, at all times. It must be located where there is good air circulation for the quick dissipation of the heat.

A hydraulic reservoir is usually provided with the following parts:
(1) Tank, (2) Baffle plate, (3) Suction line, (4) Return line, (5)
Filler-cum-breather, (6) Drain plug, (7) Strainer, (8) Fluid level
indicator, (9) Pressure gauge, (10) Removable covers, (11)
Diffuser, and (12) Magnetic tank cleaner.

Tank

A tank is used to store and condition hydraulic fluid. In order to
get better cooling and eliminate corrosion, reservoirs are made of
materials, such as mild steel or stainless steel or anodised
aluminium. Reservoir manufacturers also make them from
durable polyethylene material. They can be rectangular or
cylindrical, or 'T'-shaped or 'L'-shaped. The thickness of the steel
sheet can typically be about 2 mm.

The top cover of the tank is usually a steel plate. The bolt-
mounted cover is sealed against dust penetration. The bottom
part of the reservoir is usually inclined from side to side or 'V'-
shaped. It shall be supported on legs to a height of at least 150
mm above the floor level. The tank is also provided with a
ground connection bolt.

Suitable openings must be provided in the tank surface for fitting
clean-out plates. Gaskets should be used under all covers and
clean-out plates for tight sealing. Moreover, suction and return
lines should be fitted using flanges or heavy-duty welded
couplings. The exterior of the reservoir is usually painted, and
the interior of the reservoir is coated with rust preventive oil.
The servicing parts such as sight glasses, filters, filler breather,
and drain cock must be fitted to the tank in such a way that they
are easily accessible.

Baffle Plate

A baffle plate (Figure 2.2) is fitted lengthwise through the middle
of the tank. Its purpose is to separate the suction chamber from
the return chamber. The pump draws the fluid from the suction
chamber, and the return flow gets into the return chamber.

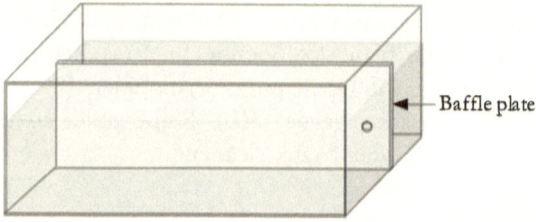

Baffle plate

Figure 2.2 | Baffle plate

The height of the baffle plate should extend slightly above the maximum fluid level in the reservoir. A small number of openings are drilled in the baffle plate at one end, far away from the suction and return lines. This provision tends to equalise the fluid levels on both sides of the baffle plate, as the fluid is directed from the return line to the suction line. It also ensures that the return fluid takes a circuitous path through the reservoir on its way from the return line to the suction line. In this way, the fluid gets more dwell time within the reservoir. The extended dwell time allows the contaminants to settle within the reservoir and assists the reservoir in dissipating the heat from the fluid as quickly as possible.

Suction Line
The suction line is used to carry the fluid from the reservoir to the inlet of the pump. Its bottom end should be located some distance above the bottom floor of the reservoir to prevent the settled contaminants from entering the pump again. The suction line is usually fitted with a strainer and/or suction filter. It may be noted that increased vacuum will develop in the suction line if the suction strainer and filter becomes clogged or the fluid remains too cold at start-up.

Return Line
The return line is used to carry the return fluid from the system back to the reservoir. The suction and return lines may be located on the same side of the reservoir, but, on either side of the baffle plate. The return line must terminate below the fluid

level and up to a height of two to four times the pipe diameter above the base plate of the reservoir to reduce the turbulence and foaming. The open end of the return line may be cut at an angle 45°, to avoid the possibility of blocking the fluid flow if the line gets pushed to the bottom of the reservoir. Further, it is advantageous to point the opening towards the sidewall of the reservoir to get as much fluid contacting surface area as possible, for faster heat transfer. The return line may be slightly over-sized or fitted with a diffuser to reduce the velocity of the return fluid.

Note: The sizing of the suction and return lines are presented in the textbook entitled 'Design of Industrial Hydraulic Systems' under the Fluid power educational series. [See Page 79, for more details]

Drain Line
A separate drain pipe can be fitted to the upper side of the reservoir for bringing the leakage fluid directly from the components through external lines without being combined with the main return flow. This provision is to prevent the back-pressure generated in the main return line from reaching the components and causing their faulty operation. The low volume, low-velocity drain flow can discharge either on top or underneath of the fluid level. The discharge of fluid above the fluid level prevents siphoning of oil out of the reservoir if an external drain connection on a component is opened.

Filler-cum-breather
An opening is usually provided at the top plate of the reservoir to act as filler-cum-breather. The opening serves as a filler to fill the reservoir with the fluid during the fluid replacement time. It also serves as a breather to pass the air in and out of the reservoir during the normal operating time. This provision tends to equalise the interior and exterior air pressures.

An air filter of 5 microns (or better), incorporated in the breather, prevents the ingress of airborne contaminants into the reservoir. The filter protects the fluid from the contamination found in the surrounding environment. It is, however, essential to install a filter of sufficient capacity to allow the rapid discharge of the air displaced by the large volume of fluid that is returning to the reservoir. The breather may include a sufficient measure of high-capacity desiccant material, like silica gel, for the dehumidification of the inflowing air. Figure 2.3 shows the cross-sectional view of an air breather.

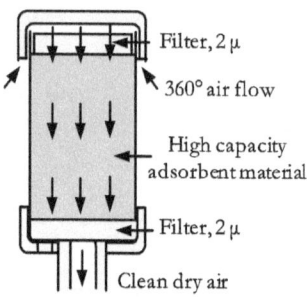

Figure 2.3 | A cross-sectional view of an air breather

It may, however, be noted that the reservoir may also be provided with a separate filler opening with a cap. The filler cap should be chained to the reservoir to keep it secured. Typical specifications of a breather are given in Table 2.1.

Table 2.1 | Typical Specifications, filler-cum-Breather

Parameter	Typical values
Diameter	94 mm or 127 mm
Maximum flow rate	280 lpm or 985 lpm
Air filtration rating	2μm, 3μm, 5 μm, 10μm
Filter element material	-Impregnated paper -Phenolic resin
Type of connection	Threaded
Tank thread connection	G¼, G 3/8, G½, G¾
Visual indicator	Yes / No

Drain Plug

A drain plug is provided at the lowest point of the bottom plate so that the system fluid can be drained entirely during the fluid replacement time. The deepest end of the reservoir is usually on the opposite side of the suction and return lines. Drain plug can also be provided on the front side of the tank. Drain plugs with hexagonal heads are typically available with different sizes from 10 mm to 250 mm. Figure 2.4 shows a drain plug.

Figure 2.4 | Drain plug

Suction Strainer

A strainer or suction filter or both are connected to the suction line to prevent dirt, grit, sludge, rust, nuts, bolts, and other large contaminants from entering the associated pump. Large particles may be introduced into the system during the assembly or maintenance of the system. The strainer is usually fitted inside the reservoir submerged in the fluid.

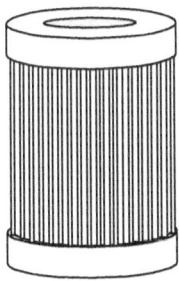

Figure 2.5 | Strainer

Figure 2.5 shows a suction strainer. The cap assembly of the strainer is epoxy bonded to the body for strength. Vacuum indicators and switches can be used along with strainers to get an indication when vacuum reaches a certain level. Table 2.2 gives the typical specifications of strainers.

Table 2.2 | Typical Specifications, strainer

Parameter	Typical values
Maximum flow rate	1140 lpm
Filter element material	Stainless steel mesh screen
Micron rating	>149μm
Fittings	Steel or Nylon
Port size	10 mm to 100 mm, Metric
Operating temperature	100°C

Fluid Level Indicator
The fluid level in the reservoir must be monitored to avoid any potential starving of the pump. This monitoring is assisted by incorporating a sight window or a fluid level indicator in the reservoir or by using a level gauge. Figure 2.6 shows the different types of fluid level indicators.

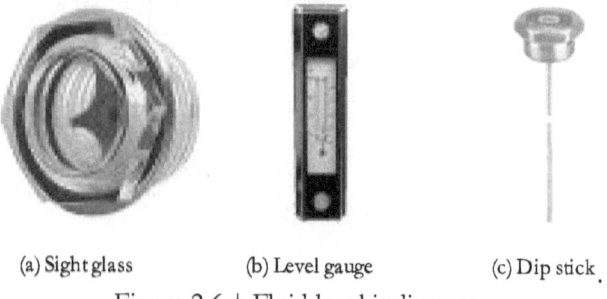

(a) Sight glass   (b) Level gauge   (c) Dip stick

Figure 2.6 | Fluid level indicators

The sight window is provided for the quick visual inspection of the fluid level in the reservoir. It comes in plastic or metal. In applications involving high pressures or temperatures, steel sight glasses are preferred. Locking nuts provide leak-free mounting into sheet metal. A transparent fluid level gauge is made of transparent lens material, such as acrylic, and is protected with a steel guard. Hexagonal-head threaded plug made of aluminium alloy with built-in dipstick of proper length can also be used to monitor the fluid level in the reservoir.

Bourdon Tube Pressure Gauge
A pressure gauge is used in a hydraulic system to measure the pressure level in the system. At the same time, the use of the pressure gauge is a safety measure as it monitors the over-pressures in the system and assists in the troubleshooting. The pump inlet condition must be checked with a vacuum gauge installed at the pump inlet.

A pressure gauge, as shown in Figure 2.7, consists of a mechanical bourdon tube with the elastic chamber, link, geared sector, pinion, pointer, calibrated scale and snubber. The bourdon tube is constructed from brass and bronze alloys. Typically, the pressure gauge has a stainless steel bezel and case which is filled with glycerin for reducing the flutter of its pointer and avoiding any internal damage.

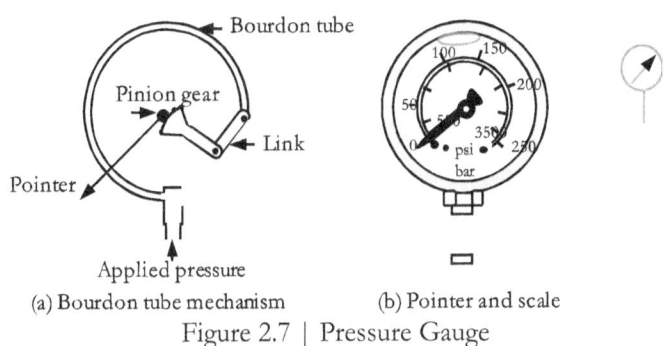

(a) Bourdon tube mechanism   (b) Pointer and scale

Figure 2.7 | Pressure Gauge

Pressure applied to the elastic chamber causes the Bourdon tube to move outwards. The higher the pressure, the greater would be the deflection radius of the tube. The deflection helps to convert the applied pressure to the corresponding movement of the attached pointer across the scale, through the links, levers, and gearing.

A fixed orifice snubber is used in the pressure connection of a gauge to dampen the pointer oscillation and protect the gauge against the damage owing to the pressure surges.

The service life of a permanently fitted pressure gauge can be prolonged by isolating the gauge from the associated system using a shut-off valve or gauge isolator, except when taking the reading. A spring-loaded pressure gauge can be used when it is desirable to monitor system pressure continuously. There is no need of a snubber or gauge isolator for this type of pressure gauge. Typical specifications of gauges are given in Table 2.3.

Table 2.3 | Typical Specifications, Pressure gauge

Parameter	Typical values
Dial sizes	63 mm, 100 mm
Scale	bar [psi]
Mounting	Stem, panel, front flange
Accuracy	±3% of the full scale
Thread type	For 63 mm: ¼ NPT, ¼ SAE, ¼ BSP For 100 mm: ½ NPT
Operating temperature	-1 to 70°C

Cleaning Cover

The reservoir must be designed for easy internal access to clean out all the residues and rust that may have accumulated in the reservoir, or for flaking paint. The periodic servicing and cleaning activities can be carried out quickly if a large opening with a removable cover (clean-out plate) is fitted on one or both ends of the reservoir. The clean-out openings should be placed to give access to both sides of the baffle plate. Figure 2.2 shows the clean-out cover.

Diffuser

Figure 2.8 shows the cut-section view of a diffuser. It is usually made of two specially-made concentric steel tubes. It is meant for reducing the velocity of the fluid returning to the reservoir.

In turn, the reduced velocity prevents foaming and the re-suspension of the deposited dirt and reduces the turbulence and noise in the system. Typical specifications of a diffuser are given in Table 2.4.

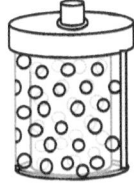

Figure 2.8 | Diffuser

Table 2.4 | Typical Specifications, Diffuser

Parameter	Typical values
Maximum flow rate	1710 lpm
Material	Perforated steel
Pipe size	12, 20, 25, 30, 40, 50, 63, 75 mm Metric
Operating temperature	Up to 120°C

Baffle Screen

A 100-mesh screen, as shown in Figure 2.9, can be installed inside the reservoir at an angle (approximately 30° from horizontal) to coalesce and collect air bubbles that can go to the top fluid surface and get dissipated.

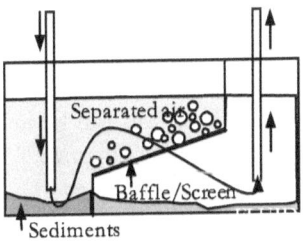

Figure 2.9 | Baffle screen

Magnetic Tank Cleaner

Abrasive ferrous particles appear in the fluid medium of a hydraulic system as a result of the constant flaking effect of its moving parts, chipping due to the sub-surface casting flaws, and particles generated out of the machining operations in the system. Tank cleaners with permanent magnets can be incorporated for attracting and holding the ferrous particles in the fluid, which ordinary filters might miss. This feature helps prevent the re-circulation of metal particles through the system and the excessive wear of the components in the system. For regular use, the standard model with a small magnetic unit and a short support rod, as shown in Figure 2.10(a), can be used. Where large-scale abuse is expected in the system operation, a multi-magnet unit with an extended support rod, as shown in Figure 2.10(b), can be used.

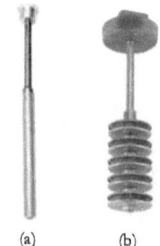

(a) (b)

Figure 2.10 | Magnetic tank cleaners

Tank Heater

If a hydraulic system is exposed to freezing temperatures, the fluid in the system reservoir becomes thick. As a result, the pump in the system is liable to be damaged due to the development of cavitation. The fluid can be kept warm with the help of an electric immersion heater (Figure 2.11).

Figure 2.11 | An electric immersion heater

Thermometer, Thermostat, and Fluid Level Switch

A thermometer measures the temperature of the fluid. A thermostat is a temperature-actuated switch that can be used for control purpose. A fluid level switch actuates a switch when the oil level reaches a particular height. The symbolic representations of these components are given in Figure 2.12.

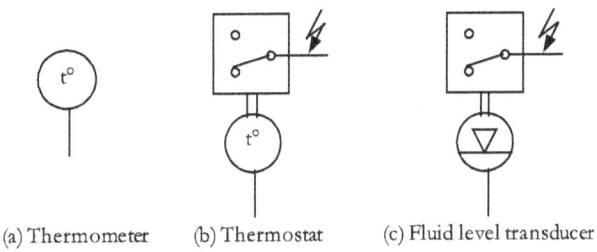

(a) Thermometer   (b) Thermostat   (c) Fluid level transducer

Figure 2.12

Pump-Reservoir Layouts

There are many methods of configuring a hydraulic power pack. The pump-motor unit can be mounted above or under or alongside the reservoir.

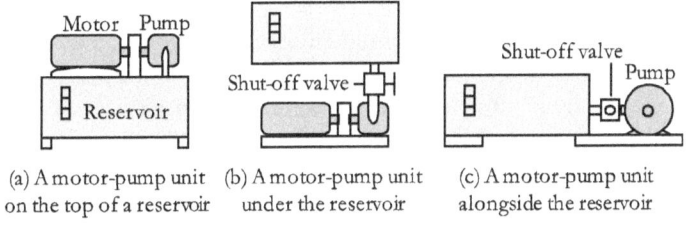

(a) A motor-pump unit on the top of a reservoir   (b) A motor-pump unit under the reservoir   (c) A motor-pump unit alongside the reservoir

Figure 2.13 | Typical layouts of hydraulic power packs

Pump-above-Reservoir

The top surface of the reservoir makes a convenient location for mounting the pump and its drive motor. Figure 2.13(a) shows the pump-reservoir layout with the pump placed above the reservoir. However, the main disadvantage of the layout is that

13

the pump must create enough vacuum at its suction side to raise and accelerate the fluid into the pump. The length of the piping must be as short as possible. The piping arrangement must have few or no bends at all.

Pump-under-Reservoir

Figure 2.13(b) shows the pump-reservoir layout with the pump unit arranged below the reservoir. With the reservoir above, the pump inlet is always flooded with the fluid. This design, therefore, improves the suction condition and reduces the possibility of whirlpools and cavitation.

Pump-alongside-Reservoir

Figure 2.13(c) shows the pump-reservoir layout with the pump, especially too heavy, alongside the reservoir. In this arrangement, the pump inlet is always flooded with the fluid. A shutoff valve in the inlet line allows the maintenance or repair work on the pump unit without draining the reservoir.

Sizing of Hydraulic Reservoirs

Reservoirs used in hydraulic systems differ in terms of their sizes (capacities). An adequately designed reservoir for a hydraulic system must be correctly sized so that it can supply the required quantity of fluid to the system at all times. Moreover, the size of the reservoir must ensure that the fluid in circulation has a reasonable dwell time in the reservoir to allow it to settle out the particulate contamination, dissipate the heat as quickly as possible, suppress any turbulence in the return fluid, accommodates the effects of thermal expansion of the system fluid, and release the entrained air. An undersized reservoir is liable to produce higher fluid temperatures than the design temperature.

The size of a hydraulic reservoir depends on the applications concerned. Determining the size of a reservoir is not an exact science. Remember, reservoirs are generally over-sized. There are two approaches to find the sizes of reservoirs. One approach is an approximate method based on the volumetric flow rate involved in the associated system. Another approach to holistically finding the most appropriate size of a reservoir is to consider the heat balance feasible in the system.

Reservoir Size, Based on the System Flow Rate

For a hydraulic system, where the mineral fluids are used, and medium-to-high frequency demands are expected, a reservoir with a capacity (cubic-metre or litres) of three to five times the volume flow rate (cubic-metre/min or lpm) of the system fluid is adequate. That is,

Reservoir size, m^3 = (3 to 5) x pump flow rate, m^3/min
Reservoir size, litre = (3 to 5) x pump flow rate, lpm

With this general rule, the fluid returned to the reservoir has three to five minutes of dwell time in the reservoir before it circulates again.

Exceptions: However, the recommended size of a reservoir should be larger than normal under certain exceptional situations. That is, a larger reservoir should be used if it is liable to be exposed to high ambient temperatures or if some special fluids are intended to be used in the system. For example, a reservoir size of five to eight times the pump flow rate per minute is recommended for a hydraulic system with a water-glycol or Polyol ester fluid medium. Further, the reservoir needs to be larger if the associated system has accumulators. However, with the mobile and aerospace hydraulic systems, the space constraints often restrict the size of the reservoirs. At the same time, with the use of a heat exchanger in a system, the reservoir size can be as small as that required to accommodate the fluid corresponding to the pump flow rate in one minute.

Standard Sizes of Reservoirs

The standard sizes of hydraulic reservoirs range from 1 to 300 litres and may go even up to 1900 litres for the custom-made models. Typical standard sizes of reservoirs include: 10, 12, 20, 30, 35, 40, 45, 50, 60, 75, 100, 150, 225, 250, and 300 litres. Typical dimensions for certain tank sizes are given in Table 2.5.

Table 2.5 | Typical tank dimensions

Tank size, litre	Dimensions (LxWxH), mm³	
	Option I	Option II
10	400x280x186	-
20	400x280x274	450x200x300
30	500x320x285	-
40	500x320x364	-
45	700x370x329	-
50	640x320x400	-
60	700x370x394	600x470x485
100	700x550x565	810x560x418
250	1006x610x680	-

Reservoir Size, Based on the Heat Balance

For finding the heat balance in an existing system or in a system that is being designed, it is necessary to find the amount of heat generated as a result of energy losses taking place in the system, and then to determine how much heat can be dissipated from the heat-dissipating area of the reservoir. From these calculations, the optimum size of the reservoir and the size of the heat exchanger can be determined.

Heat Load of an Existing Reservoir

Consider an existing hydraulic system with a reservoir of known volume. The fluid in the system gets heated up owing to the inefficiencies in the system. Devices used in the hydraulic system, such as the flow control valves, sequence valves, pressure reducing valves, and undersized directional control valves can contribute to the development of heat in the system. The fluid temperature tends to increase initially as the system is switched

16

on until a levelling off temperature is reached. The heat load of the reservoir with a fluid can easily be determined, using the following procedure:

1. Measure the volume of the reservoir (V, litre)
2. Measure the fluid temperature (T1) at the start-up
3. Measure the fluid temperature (T2) after the system has been in operation for a specified duration (Δt, in minutes)
4. Determine the differential temperature, $\Delta T = T2 - T1$

$$\text{Heat load (KW)} = \frac{V \text{ (litre) x } \Delta T \text{ (°C)}}{32.4 \text{ x } \Delta t \text{ (min)}}$$

Heat Load of a New Reservoir being Designed
The heat load of a new reservoir being designed can be found by using the input power to the system as a guide. It can be obtained from the nameplate of the drive motor or engine. The losses of the components of a system can be taken from the manufacturer's published data. Generally, the losses added by pumps and actuators can be taken as 5 to 15%, and that by valves and flow restrictions can be taken as 10 to 20% of the input power involved in every component. The total heat losses in a hydraulic system would be about 30 to 50% of the input power of the drive motor or engine.

Heat dissipation by Hydraulic Reservoirs
The heat-dissipation factor, which is an essential consideration in deciding the reservoir capacity, is explained below. Assume that a reservoir with a heat-dissipating surface area 'A' dissipates heat 'H' through conduction and radiation. The base of the reservoir can be excluded from the heat-dissipating surface if it is not elevated by at least 150 mm from the ground. Let ΔT be the temperature difference between the reservoir walls and the ambient air. The formulae for the heat dissipated by the reservoir are given by:

$H = k \times \Delta T \times A$, where k is a constant (General formula)
$H \text{ (kW)} = 0.016 \times \Delta T \text{ (°C)} \times A \text{ (m}^2\text{)}$

Example 2.1

If the fluid temperature in a hydraulic reservoir is 60°C, and the ambient air temperature is 24°C, how much heat will the reservoir with 1.858 m² of surface area dissipate?

Solution

Fluid temperature	= 60°C
Ambient air temperature	= 24°C
Temperature differential	= 60 – 24 = 36°C
Surface area of the reservoir	= 1.858 m²

Heat dissipation, H
$$= 0.016 \times \Delta T \, (°C) \times A \, (m^2)$$
$$= 0.016 \times 36 \times 1.858 = 1.07 \, kW$$

Example 2.2

A steel tank with dimensions of 500mmx400mmx420mm designed for use as a hydraulic reservoir dissipates heat from the hydraulic fluid at 70 °C. The ambient temperature is 25°C. Calculate the heat-dissipation rate of the tank.

Solution

Tank dimension	= 500mmx400mmx420mm
Fluid temperature	= 70°C
Ambient temperature	= 25°C

$$\Delta T = 70 - 25 = 45°C$$

$$A = (2 \times 500 \times 420 + 2 \times 400 \times 420 + 500 \times 400) \times 10^{-6} \, m^2$$
$$= (420000 + 336000 + 200000) \times 10^{-6} \, m^2$$
$$= 956000 \times 10^{-6} \, m^2$$

Note: The base area of the reservoir is excluded from the heat-dissipating surface, assuming that the base surface is not sufficiently elevated from the ground.

$$H = 0.016 \times \Delta T \times A \, (m^2)$$
$$= 0.016 \times 45 \times 956000 \times 10^{-6} = 0.68832 \, kW$$

Heat Exchangers

If the cooling effect from the reservoir is insufficient, a heat exchanger (or cooler) must be fitted to increase the heat dissipation rate. However, the heat exchangers are expensive, and the maintenance of them can run high. Two main types of heat exchangers are used in hydraulic systems. They are: (1) air-cooled heat exchangers and (2) water-cooled heat exchangers.

Air-cooled Heat Exchanger

When the heat to be removed in a hydraulic system is comparatively small, an air-cooled heat exchanger can be employed. The schematic diagram of the typical air-cooled heat exchanger is shown in Figure 2.14.

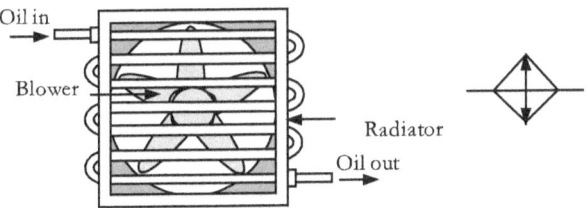

Figure 2.14 | Air-cooled heat exchanger

It primarily consists of an electric-motor-driven blower and radiator. The fixed-speed or variable-speed blower forces air across the radiator to cool the warm fluid flowing through the radiator. The motor speed can be varied using an open-loop or closed-loop PID controller and temperature sensors. The fan continually consumes energy and creates permanent noise. The airborne contaminants in the atmosphere surrounding the blower, such as heavy dust, water, and coolant vapours can quickly reduce the efficiency of the air-cooled heat exchanger.

Water-cooled Heat Exchanger

When the heat to be removed is comparatively high, or its surrounding atmosphere is liable to be hot, a water-cooled heat exchanger should be used in a hydraulic system. Figure

2.15 shows the schematic diagram of the commonly used shell-and-tube type of water-cooled heat exchanger. It may be noted that the shell is usually made of brass. It surrounds a series of small tubes. The fluid flows through the tubes, and cold water flows through the shell surrounding the tubes. The heat from the fluid is carried away by the relatively cold water.

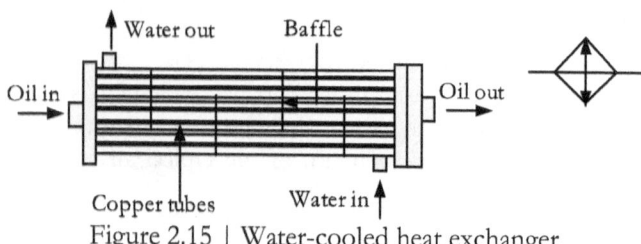

Figure 2.15 | Water-cooled heat exchanger

The essential specifications of air-cooled heat exchangers and water-cooled exchangers are given in Table 2.6 and Table 2.7, respectively.

Table 2.6 | Typical specifications, air-cooled heat exchanger

Parameter	Typical values
Type of fluid	Mineral, fire-resistant, biodegradable
Fluid flow rate	13 to 300 lpm
Fluid operating viscosity	130 to 500 cSt
Air flow	650 to 8200 m³/hr
Power, fan motor	0.03 to 3 kW @ 50 Hz
Motor speed	1000, 1500, 3000 rpm
Noise level, at 1 m distance	60 to 84 dB(A)

Table 2.7 | Typical specifications, water-cooled heat exchanger

Parameter	Typical values
Cooling capacity	10, 20, 30, 100 kW
Flow rate	5 to 100 lpm
Pressure, maximum	5.5 bar
Reservoir capacity	7 to 70 litre

Chapter 3 | Hydraulic Pumps

A hydraulic power pack should be provided with the operational pump-motor unit. It may also be provided with a standby pump-motor unit. The pump converts the mechanical energy from the motor into hydraulic energy, using an incompressible fluid medium.

It draws the fluid from the associated reservoir when driven by its prime mover and then pushes it into the system. This pumping action creates flow. The system develops pressure owing to the resistance offered to the flow. The resistance to the flow arises due to the fluid viscosity, flow restrictions, and/or the load on actuators.

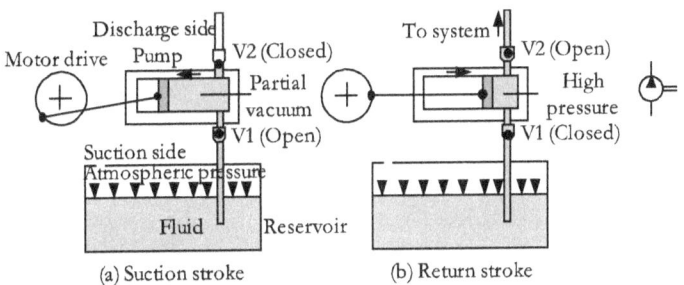

(a) Suction stroke   (b) Return stroke

Figure 3.1 | Schematic diagrams showing two critical positions of a hydraulic pump system

The pumping action is explained with the help of the schematic diagram of a simple piston pump, as shown in Figure 3.1. The piston moves back and forth when driven by a motor at a constant speed and acts as the pumping element.

The pump incorporates two valves V1 and V2 for drawing and containing the fluid within the system.

The suction stroke creates a partial vacuum at the suction side of the pump and draws the fluid from the associated tank into the pump chamber through the opened valve V1. On the return stroke, the valve V1 is closed, and the fluid is trapped in the chamber. The closed valve V1 provides tight sealing and prevents the fluid from flowing back to the tank.

The motion of the pumping element induces a force to the pumped fluid and discharges the fluid through the opened valve V2. In this way, the pump creates a flow. In short, during the pumping action, the pump draws, traps, seals, and discharges the fluid.

As stated earlier, a partial vacuum is created at the suction side of a pump. If a high vacuum is developed at the pump inlet, it can cause the excessive wear of the internal surfaces due to cavitation. Therefore, the pump should be located at an elevation of not more than 180 cm above the fluid level in the reservoir. Practically, the height should be limited to 90 cm, as other restrictions in the inlet line will add to the vacuum.

Further, it must be the design goal to construct a pump that runs quietly without producing vibrations. The use of special seals and the provision of close tolerance characterise the construction of the positive-displacement pump. Under normal pressures, these features assist in transferring a fixed volume of fluid during each cycle with no leakage, thus giving rise to the term 'positive displacement'. However, the pump is liable to promote leakage with the increased friction of the pumping elements.

A pump should be constructed to withstand high pressures, create low flow pulsations, and produce low noise levels.

Terms and Definitions -Hydraulic Pumps

The positive-displacement pumps come in many different varieties, sizes, flow rates, and power ratings. Some important parameters concerned about the operation of hydraulic pumps are its pressure rating, priming, slippage, displacement, flow rate, torque, input power, output power, and efficiency.

Pressure Rating

It is the pressure that overcomes all resistances in the system, which includes both useful work and losses. Alternatively, it is the maximum pressure the pump is capable of withstanding without any damage to its parts or any excessive increase in its internal leakage. If an application involves simple or moderate work, a low to medium pressure pump would be the most suitable. On the other hand, if an application requires substantial work, as in large construction equipment, a high-pressure pump would be the most appropriate.

Volumetric Displacement (V_D)

It is the volume of the fluid that is carried by the pump in one revolution of its driveshaft. It is expressed in cubic-centimetre per revolution (cc/rev) or litres per revolution, cubic-metre per revolution (m^3/rev), or other similar units

Theoretical Flow Rate (Q_T)

It is the volume of the fluid displaced by the pump, at its inlet, per unit of time. It can be determined by the product of the volumetric displacement of the pump and the speed of the pump's driveshaft. In the SI system of units, the theoretical flow rate is commonly measured in cubic metre per second (m^3/s) or lpm.

The mathematical equation for the theoretical flow rate (Q_T) of the pump in the SI system of units is as follows:

$$Q_T \ (m^3/min) = V_D \ (m^3/rev) \ x \ N \ (rpm)$$

Pump Slippage (Q_s)

It represents the internal leakage of the fluid in the pump from its discharge port to its suction port. The internal leakage is due to some unavoidable small clearance that exists between the internal parts of the pump. The slippage is a function of the pump speed, differential pressure across the pump, degree of wear of its interior surfaces, and viscosity of the fluid passing through the pump. Any increase in the slippage leads to lesser efficiency of the pump.

Actual Flow Rate (Q_A)

It is the actual fluid volume discharged by the pump per unit of time. It is given by the theoretical flow rate minus the pump slippage. That is,

Actual flow rate = Theoretical flow rate – Slippage

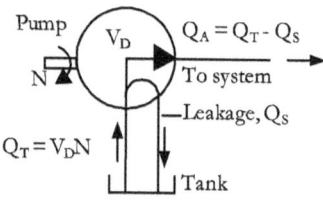

Figure 3.2 | A symbolic representation of a hydraulic pump showing its flow rate parameters

In the SI system of units, the actual flow rate is commonly measured in cubic metre per second (m^3/s) or lpm. Figure 3.2 depicts the relationship between the theoretical and actual flow rates in the pump drawing fluid from the tank and delivering to the system.

Example 3.1
What is the theoretical flow rate of a fixed-displacement pump with a volumetric displacement of 0.131x10⁻³ m³/rev operating at 2000 rpm?

Solution
Volumetric displacement, V_D = 0.131x10⁻³ m³/rev
Pump speed, N   = 2000 rpm = 33.33 rps

Theoretical flow rate, Q_T = V_D x N
                           = 0.000131x33.33
                           = 0.00436 m³/s

Actual Torque (T_A)
It is the actual torque delivered to the pump by its prime mover and is given by:

$$T_A(Nm) = \frac{\text{Actual power delivered to the pump (watt)}}{\omega \ (rad/s)}$$

Theoretical Torque (T_T)
The theoretical Torque of the pump is a function of the volumetric displacement of the pump and the system pressure. It is equal to the actual torque minus the torque losses on account of the friction in the pump.

$$T_T(Nm) = \frac{P(Pa) \times Q_T (m^3/s)}{\omega(rad/s)} = \frac{P \times (V_D \times n)}{(2\prod \times n)}$$

That is,

$$T_T(Nm) = \frac{V_D \ (m^3/rev) \times P(Pa)}{2\prod}$$

Pump Input Power

It is the power delivered to the pump by its prime mover. The speed of the driveshaft and torque imparted by the motor determine the input power to the pump.

$$\text{Pump input power (kW)} = \frac{T_A \text{ (Nm) x N (rpm)}}{9550}$$

$$\text{Pump input power (kW)} = \frac{T_A \text{ (Nm) x } \omega \text{ (rad/s)}}{1000}$$

Pump Output Power

It is the power delivered by the pump. The pressure and the actual flow rate of the pump determine the output power.

$$\text{Pump output power (kW)} = \frac{P \text{(Pa) x } Q_A{}^{(m^3/s)}}{1000}$$

$$\text{Pump output power (kW)} = \frac{P \text{(bar) x } Q_A \text{(lpm)}}{600}$$

Example 3.2

A hydraulic pump delivers 40 lpm at 150 bar for carrying out a work operation. Calculate the hydraulic power developed by it.

Solution

With usual notations,

Flow, Q_A	= 40 lpm
Pressure, P	= 150 bar

Output power, P_{out} = P (bar) x Q_A (lpm) / 600

$$= 150 \text{ x } 40 / 600 = 10 \text{ kW}$$

Efficiencies of Hydraulic Pumps
In a practical hydraulic pump, both fluid leakage and frictional loss take place and, therefore, their efficiency is always less than 100%. Two types of efficiencies are identified to account for the two types of losses in the pump. They are: (1) Volumetric efficiency, and (2) Mechanical efficiency.

Volumetric Efficiency (η_v)
It represents the ratio of the actual pump flow rate at a given pressure and the theoretical flow rate as determined by the geometric displacement of the pump, assuming no frictional losses in the pump. It indicates the extent of leakage that takes place within the pump. It is given by:

$$\text{Volumetric Efficiency} \left(\eta_v \right) = \frac{\text{Actual flow rate}}{\text{Theoretical flow rate}} = \frac{Q_A}{Q_T}$$

Mechanical Efficiency (η_m)
It represents the ratio of the power delivered by the pump to the power delivered to the pump, assuming no leakage in the pump. It indicates the amount of energy losses that take place in the pump due to friction. It is given by:

$$\text{Mechanical Efficiency} \left(\eta_m \right) = \frac{\begin{array}{c}\text{Pump output power,}\\ \text{assuming no leakage}\end{array}}{\begin{array}{c}\text{Actual power delivered}\\ \text{to the pump}\end{array}}$$

$$\text{Mechanical Efficiency} \left(\eta_m \right) = \frac{P \text{x} \, Q_T}{T_A \text{x} \, N}$$

The mechanical efficiency of the pump can also be calculated in terms of its torque units. That is,

$$\text{Mechanical Efficiency} \left(\eta_m \right) = \frac{T_T}{T_A}$$

Overall Efficiency (η_o)

It is the ratio of the actual power delivered by the pump to the actual power delivered to the pump. It is given by:

$$\text{Overall Efficiency } \left(\eta_o\right) = \frac{\text{Actual power delivered by the pump}}{\text{Actual power delivered to the pump}}$$

$$\text{Overall Efficiency } \left(\eta_o\right) = \frac{P \times Q_A}{T_A \times N}$$

The overall efficiency is also given by the product of its volumetric efficiency (η_v) and mechanical efficiency (η_m). That is,

$$\eta_o = \eta_v \times \eta_m$$

Summary of Relations for Hydraulic Pumps

Figure 3.3 gives the summary of essential relations of hydraulic pumps for the easy understanding and the correlation of these relations by the reader.

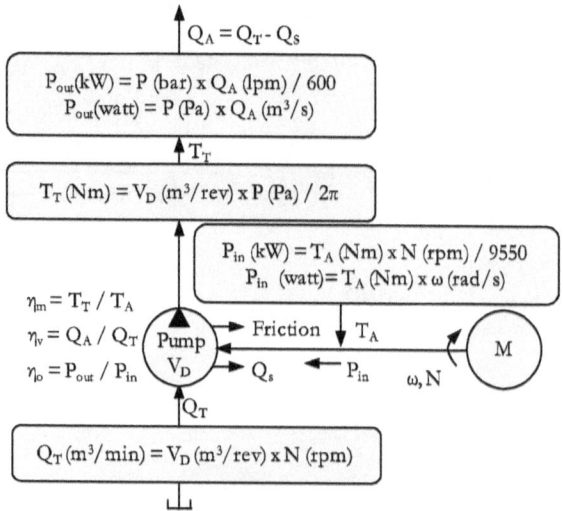

Figure 3.3 | Summary of relations for hydraulic pumps

Example 3.3

A hydraulic pump with a displacement of 104 cc/rev delivers 1.6×10^{-3} m³/s operating at a pressure of 100 bar when driven by a prime mover with a torque of 180 Nm at 1000 rpm. Find out: (1) the volumetric efficiency, (2) the mechanical efficiency, and (3) the overall efficiency, of the pump.

Solution

Pump displacement, V_D	= 104 cc/rev
Actual flow rate, Q_A	= 1.6×10^{-3} m³/s
Speed, N	= 1000 rpm
Pressure, P	= 100 bar
Actual torque, T_A	= 180 Nm

Theoretical flow rate, $Q_T = V_D$ (m³/rev) x N (rev/s)
$$= 104 \times 10^{-6} \times (1000/60)$$
$$= 1.73 \times 10^{-3} \text{ m}^3/\text{s}$$

Volumetric efficiency, $\eta_v = (Q_A/Q_T) \times 100\%$
$$= (1.6 \times 10^{-3}/1.73 \times 10^{-3}) \times 100\% = 92\%$$

Input power P_{in} = T_A(Nm) x N(rpm)/9550
$$= 180 \times 1000 /9550 = 18.8 \text{ kW}$$

Output power, P_{out} = P(Pa) x Q_A(m³/s)/1000
$$= (100 \times 10^5 \times 1.6 \times 10^{-3})/1000$$
$$= 16 \text{ kW}$$

Output power assuming no leakage = P x Q_T/1000 kW
$$= (100 \times 10^5 \times 1.73 \times 10^{-3})/1000$$
$$= 17.3 \text{ kW}$$

Mechanical efficiency, $\eta_m = (P_{out, \text{No leakage}}/ P_{in}) \times 100\%$
$$= (17.3/18.8) \times 100\% = 92\%$$

Overall efficiency, $\eta_o = \eta_v \times \eta_m = 0.92 \times 0.92 = 0.846 = 85\%$

Classification and Types of Hydraulic Pumps
In general, hydraulic pumps are broadly classified into the following two types. They are: (1) positive-displacement (hydrostatic) pumps and (2) non-positive-displacement (hydrodynamic) pumps.

The positive-displacement pump delivers a definite amount of fluid to the system. In the non-positive-displacement pump, its internal leakage increases as the downstream pressure increases. A classification chart for hydraulic pumps is given in Figure 3.4.

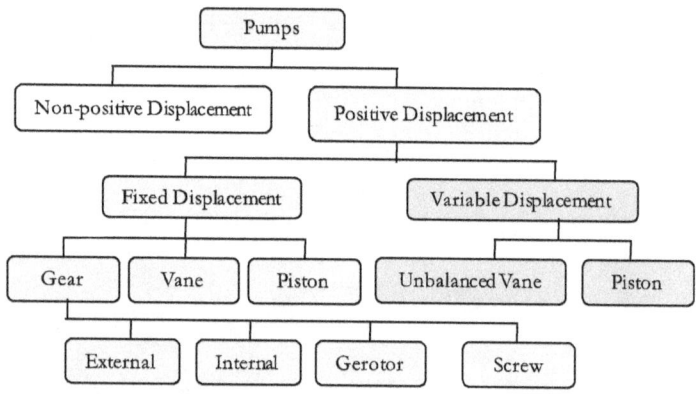

Figure 3.4 | Classification chart for hydraulic pumps

The positive-displacement pumps can be classified into the fixed-displacement type and variable-displacement type. The fixed-displacement pump delivers a fixed amount of fluid for each revolution of its driveshaft. In contrast, the variable-displacement pump is designed to deliver a variable volume of fluid per cycle when running at a constant speed. The pump output, in this case, can be varied by changing the physical relationship of the internal pump elements. However, variable-displacement pumps are more involved in the design aspect.

The positive-displacement pumps can further be classified according to the types of pumping elements, such as gears, vanes, and pistons, used. Accordingly, there are three main types of positive-displacement pumps. They are: (1) Gear pumps, (2) Vane pumps and (3) Piston pumps.

Symbolic Representation of Hydraulic Pumps
Figure 3.5 gives the symbols of various types of hydraulic pumps.

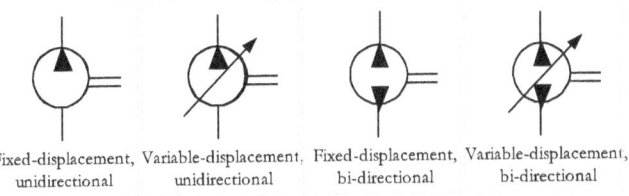

Fixed-displacement, unidirectional   Variable-displacement, unidirectional   Fixed-displacement, bi-directional   Variable-displacement, bi-directional

Figure 3.5 | Symbols of variants of hydraulic pumps

Gear Pumps
Gear pumps consist of two or more gears meshing with each other. They always act as fixed-displacement pumps. Gear pumps are classified into: (1) External gear pumps, (2) Internal gear pumps, (3) Gerotor pumps, and (4) Screw pumps.

External gear pump: It consists of two close-meshing identical gears (i.e., gear-on-gear), enclosed in a close-fitting housing with side wear plates.

Internal gear pump: It consists of an outer rotor gear, an inner spur gear, and a crescent-shaped spacer, all enclosed in a housing.

Gerotor pump: It consists of an off-centre inner gear rotor (Gerotor element) and an outer female gear rotor (idler) enclosed within the housing of the pump.

Screw pump: It is constructed with two parallel intermeshing rotors (screws) capable of rotating in a housing machined to close tolerances.

Some of the most important advantages and disadvantages of gear pumps are given in Table 3.1.

Table 3.1 | Advantages and disadvantages of gear pumps

Gear pump type	Advantages	Disadvantages
External gear pump	Simple, rigid, more tolerant of fluid contamination, least expensive	Unbalanced, produce more wear and leakage, greater fluid pulsations
Internal gear pump	Compact, reliable, provides smooth non-pulsating flow, good suction ability, produce less cavitation,	More complicated design, expensive to manufacture, overhung load on shaft bearings
Gerotor pump	Design flexibility	Limited to medium pressures, excessive fluid leakage
Screw pump	Good suction characteristics, high tolerance to contamination, produce non-pulsating delivery	Complicated screw shape, low efficiency, high cost

Vane Pumps

According to the type of design, vane pumps can be classified into two types. They are: (1) unbalanced vane pumps and (2) balanced vane pumps. They can also be classified into fixed displacement type and variable-displacement type.

Unbalanced Vane Pump: It consists of a prime-mover-driven rotor with sliding vanes in close-fitting radial slots.

Variable-displacement Vane Pump: The displacement of unbalanced vane pump can be varied by modifying the degree of eccentricity between the rotor and the casing of the pump from zero to some maximum.

Balanced Vane Pump: It consists of a prime-mover-driven rotor with sliding vanes in close-fitting radial slots. The rotor moves inside an elliptical cam ring (casing) concentrically. The pump also consists of a set of diagonally opposite suction ports as well as discharge ports. This design creates two separate pumping areas on opposite sides of the rotor.

Some of the most important advantages and disadvantages of vane pumps are given in Table 3.2.

Table 3.2 | Advantages and disadvantages of vane pumps

Advantages	Disadvantages
Excellent suction characteristics, low pulsation, low noise level, work well with low-viscosity fluids, support both fixed- and variable-displacement deliveries	Complex structure, more expensive than gear pumps, sensitive to contamination, uneven wear of sliding contacts, unsuitable for high pressures, require fluid with excellent anti-wear properties

Piston Pumps

A piston pump consists of a cylinder block with pistons attached to the drive shaft. The piston pumps can broadly be classified as axial-piston pumps and radial piston pumps according to the spatial arrangement of the associated cylinders. In an axial-piston pump, the pistons are arranged parallel to the cylinder block, whereas, in a radial piston pump, the pistons are arranged radially in the cylinder block. The piston pumps are also available in the fixed-displacement and the variable-displacement designs.

Axial-piston Pumps: It consists of many pistons arranged in a circular array within the associated cylinder block. The cylinder block assembly rotates with the drive shaft. The axial-piston pumps can further be classified as in-line axial-piston pumps and bent-axis axial-piston pumps.

In-line Axial-piston Pumps

It mainly consists of (1) a group of cylinders with pistons, (2) a cam plate (swash plate), (3) a stationary valve plate, and (4) a drive shaft.

The cylinders are arranged in parallel and formed into a round block about an axis. The axis of the cylinder block is aligned in-line with the axis of the drive shaft. The cylinder pistons are fitted to the cam plate through ball joint shoes.

If the cam plate is not angled, the rotating pistons tend to remain stationary within the bores. However, if the cam plate is angled, the rotating pistons move back and forth in their respective cylinder bores.

The valve plate contains two kidney-shaped openings. The larger opening is the suction port, and the smaller opening is the discharge port.

The bores in the cylinder block are connected successively to the low-pressure suction port and the high-pressure discharge port by the valve plate while rotating. This action creates the flow.

Bent-axis Piston Pump

It mainly consists of a: (1) cylinder block, (2) driveshaft, and (3) valve plate.

The axis of the cylinder block is set at an angle to the axis of its drive shaft through a jointing mechanism that tends to reduce the side loads. When driven, the pistons reciprocate in their respective bores. The valve plate contains two semi-circular openings for the suction port and the discharge port.

The bores in the cylinder block are connected successively to the low-pressure suction port and the high-pressure discharge port by the valve plate while rotating. This action creates the flow.

Radial Piston Pumps

This design consists of (1) a cylinder block with seven or nine radial barrels, (2) a reaction ring, (3) a pintle, and (4) a drive shaft.

The pistons are arranged in radial directions around the drive shaft. The pintle includes suction and discharge ports. The pistons in the cylinder block remain in constant contact with the reaction ring due to the centrifugal force and the back pressure on the pistons. Next, the rotor moves eccentrically to the cylinder block. Further, the pistons move in and out, as the cylinder block rotates.

The inner openings of the cylinder barrels are connected successively to the low-pressure suction port and the high-pressure discharge port, while rotating, to direct the fluid in and out of each cylinder. Some of the most important advantages and disadvantages of piston pumps are given in Table 3.3.

Table 3.3 | Advantages and disadvantages of piston pumps

Advantages	Disadvantages
High power-to-weight ratio	Intricate design
High efficiency	Most expensive
Highly reliable	High maintenance cost
Operate at high pressures	Sensitive to contamination

Appendix 1 gives the typical pump data extracted from the manufacturer's domain.

Comparison of Hydraulic Pumps

Table 3.4 provides the comparison of various types of positive-displacement pumps.

Table 3.4| Comparison of positive-displacement pumps

Parameter	Gear pumps	Vane pumps	Piston pumps
Design	Simple, rugged	Slightly complex	Complex
Displacement type	Fixed	Fixed or variable	Fixed or variable
Pulsation	High	Low	High
Fluid sensitivity	Least sensitive	Sensitive	Sensitive
Effect of contaminants	Tolerant	Less tolerant	Sensitive
Leakage	Prone to leakage	Less prone	Prone to leakage
Noise level	High	Low	High
Size	Small to medium	Small to medium	Medium to large
Power-to-weight ratio	Low	Low	High
Efficiency	Least	Medium	Highest (>90%)
Cost	Cheapest	Medium	Costly
Maintenance costs	High, due to wear	High, due to wear	Very high
Service life	Longest	Long	Very long
Application	Light-, medium-duty	Light-, medium-duty	Heavy-duty
Pressure level	Medium	Lowest	Highest
Displacement	Medium	Lowest	Highest
Viscosity range	highest	Low	Low

Mounting of Hydraulic Pumps

By and large, hydraulic pumps are mounted horizontally on the top plates of the reservoirs. They, often employ foot or flange mounting designs.

The mounting arrangements in a pump are often provided at the end caps of the pump. That is; flanges are provided for the stationary mounting, and clevises-type fittings are provided for the swing type mounting. The mounting flange should remain perpendicular to the drive shaft.

The pump's shaft and the prime mover's shaft must be properly aligned at the time of their assembly. The maximum permissible parallel misalignment is, usually, less than 0.1 mm, and the maximum permissible angular misalignment is, usually, less than 0.2°, or according to the recommendations of pump manufacturers.

Employ, a flexible coupling, whenever possible, to ensure optimal damping. Avoid any stress on the shaft against the bending or thrust loads. Further, the mounting bolts should be properly seated and torqued.

The direction of rotation of a uni-directional hydraulic pump must be the same as that indicated on the pump.

Side Loads on Hydraulic Pumps

In certain hydraulic pumps, such as gear pumps and unbalanced vane pumps, all of the pumping action of the high-pressure fluid takes place in the chambers on one side of their rotors and drive shafts. This type of design imposes a considerable amount of side loads on the rotors and the shafts. A misaligned pump shaft can also impose large side loads on its driveshaft. The side loads in a pump can cause bearing failure, seal damage, and eventual shaft breakage.

Characteristic Curves of Hydraulic Pumps

The performance of hydraulic pumps can be assessed by using the following characteristic curves (1) Flow Vs Speed, (2) Flow Vs Operating pressure, (3) Drive power Vs Operating pressure and (4) Efficiencies Vs Operating pressure.

The flow rate curve, as shown in Figure 3.6(a), shows the relation of the flow rate of a hydraulic pump with the speed of its drive shaft, at a specified pressure. Figure 3.6(b) shows a graph of the pressure versus flow rate. Ideally, there is no internal slippage in a positive-displacement pump. However, in reality, the mating components are not a perfect fit, and small leaks do occur past the clearances inside the pump. The leak increases with increase in pressure. The self-explanatory Figures 3.6(c) and 3.6(d) give the pressure versus power and the pressure versus efficiency curves.

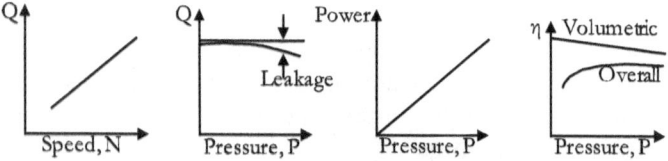

(a) Flow Vs Speed (b) Pressure Vs Flow (c) Pressure Vs Power (d) Pressure Vs η

Figure 3.6 | Typical characteristic curves of hydraulic pumps

Requirements of Hydraulic Pumps

The most critical customer requirements, when it comes to selecting pumps, are their compact dimensions, minimal pressure and volume pulsations, high efficiency, low cost, long service life, reduced vibration, and low operating noise levels.

Selection of Hydraulic Pumps

Hydraulic pumps of varying designs and types are available in the market. The designer must satisfy many application and system requirements before selecting the types of pumps for the system. Some of the application requirement parameters are the expected load, type of fluid and its viscosity, operating temperature range, overall cost, and permissible noise level. Some of the system requirement parameters are the maximum pressure, flow rate, displacement, duty cycle, reliability, and ease of maintenance.

Application Notes, Hydraulic Pumps

Various types of hydraulic pumps are available to meet the wide-ranging application demands in industrial, mobile, aerospace, marine, mining, agriculture, and construction fields.

For light-duty and medium-duty industrial applications, gear pumps and vane pumps are most used, whereas, for power-intensive applications, piston pumps are most suitable. For applications involving rough handling and dirty environment, as in mobile equipment and conveyor systems, external gear pumps are most appropriate. Internal gear pumps find applications requiring low-speed and quite operations, as in hydraulic presses, drilling machines, and lifting devices, and in marine and petrochemical applications. For energy-efficient applications, where space and weight is a premium, as in aircraft, Gerotor pumps can be used. Screw pumps find applications involving quiet operations, as in machine tools, hydraulic presses, rolling mills, sheet metal machines, plastic moulding machines, hydraulically-driven propellers, submarines, and off-line filtration systems.

For sophisticated applications involving variable-displacement and low-noise operations, vane pumps are suitable. They are found in automotive power steering and transmission applications, marine and railway winches, oil field and drilling equipment, earthmoving and construction equipment, plastic injection moulding machines, machine tools, and large presses.

Piston pumps find applications in aerospace, agricultural, automotive, mobile, and construction equipment, marine equipment, metal forming and stamping machines, machine tools, oilfield equipment, and mining fields.

Chapter 4 | Pressure Relief Valves

A pressure relief valve prevents the over-pressurisation of a hydraulic system and damage to pump, hoses and filters in the system. It modulates the flow through a hydraulic system to keep its working pressure at the preset level. It is designed to re-close after the normal pressure condition has been restored. A drain path is provided to relieve any leakage fluid.

The required system pressure (Ps) can be set on the PRV. When the actual pressure in the system is less than the set pressure, the flow is directed to the system. When the actual pressure in the system is higher than the set pressure, the flow is diverted to the reservoir.

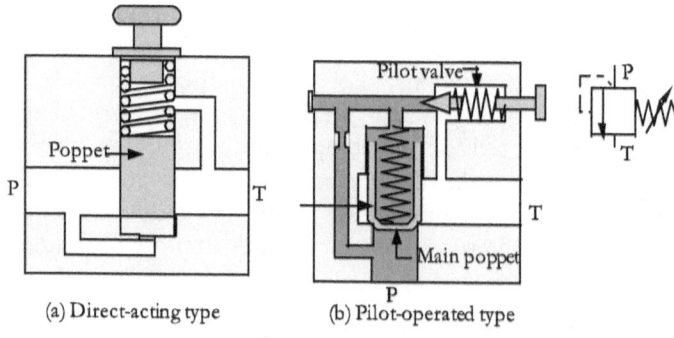

(a) Direct-acting type   (b) Pilot-operated type

Figure 4.1 | Pressure relief valve

Direct-acting PRVs [Figure 4.1(a)] have large pressure overrides and lower precision. They are used for systems with small values of pressure and flow.

Pilot-operated PRVs [Figure 4.1(b)] can be set to open completely, over a narrow range of pressures. They can pass large flows through it with a minimum pressure buildup. They provide higher precision than that of direct-acting PRVs.

Table 4.1 gives the most important specifications of pressure relief valves.

Table 4.1 | Typical Specifications, Pressure relief valve

Parameter	Typical values
Type	Direct-acting or Pilot-operated
Pressure adjustment	Slotted head screw with a lock nut or Hand knob
Valve nominal size	NG 06, 10, 16, 25, 40, 50, 80
Connection	Cartridge, threaded, sub-plate, ISO mounting
Maximum pressure	Up to 630 bar
Fluid viscosity	Recommended; 30 to 80 cSt Permitted: 20 to 380 cSt
Fluid cleanliness	ISO 4406: 18/16/13
Fluid temperature	-20 to +60°C

Sizing of a PRV

The proper sizing of a PRV used in a hydraulic system is most important to obtain the optimal protection of the system. The PRV must be selected according to the following factors. They are: (1) the fluid properties, such as viscosity, specific gravity, and compressibility, and (2) operating conditions, such as set pressure, temperature, relieving capacity (lpm), and allowable percentage of over-pressure. The following paragraph gives the empirical formulae for calculating the effective flow path area of the spring-loaded and the pilot-operated PRVs. The appropriate valve size and style may then be selected having a nominal effective area equal to or greater than the calculated area.

However, it may be noted that the empirical formulae are given only for educational purpose and may not be sufficient for deriving the sizes of PRVs in complex systems. The reader is advised to refer to the industry standards, and/or use the approved design tools for finding the design parameters for PRVs intended for industrial hydraulic systems.

For Direct-acting pressure relief valves

$$A = \frac{15.9\, Q\sqrt{SG}}{K_W K_V \sqrt{\Delta P}}$$

For Pilot-operated pressure relief valves

$$A = \frac{12.16\, Q\sqrt{SG}}{K_V \sqrt{\Delta P}}$$

Where:

A = Minimum active discharge area of the orifice, mm^2
SG = Specific gravity of the liquid at flowing conditions
Q = Relieving capacity at flowing temperature lpm
ΔP = Differential pressure (=set pressure + overpressure – backpressure) kPa
Kv = Viscosity correction factor
Kw = Backpressure correction factor for direct-acting PRVs. (For all other valves Kw = 1)

Exercise 4.1

Estimate the active discharge area needed for a pilot-operated pressure PRV to relieve 475 lpm of the fluid. The specific gravity of the fluid medium is 1.23. The set pressure is 6.9 bar, backpressure 0 - 2.07 bar. The over-pressure is 10%. Take Kv = 1.0 for the fluid.

Solution

Flow, Q	= 475 lpm
Set pressure, P_s	= 690
Over-pressure, P_o	= 10% of 690 kPa = 69
Backpressure, P_b	= 2.07 bar = 207 kPa
SG	= 1.23
Kv	= 1.0
ΔP	= $P_s + P_o - P_b$
	= 690 + 69 – 207 = 552 kPa
Active discharge area, A	= $(12.16\, Q\sqrt{SG})/(Kv\sqrt{\Delta P})$
	= $(12.16 \times 475\sqrt{1.23})/(1.0 \times \sqrt{552})$
	= 273 mm^2

A valve with the next standard size should be selected.

Chapter 5 | Hydraulic Valves

Hydraulic valves are used in a hydraulic system to realise the directional and speed control requirements of actuators in the system. Accordingly, hydraulic valves are classified into directional controls valves, flow control valves, and pressure control valves.

A directional control valve can be used to control the direction of motion of an actuator connected to the system. A non-return valve (NRV) allows the flow of the pressurised fluid in a hydraulic system in only one direction and blocks the flow in the opposite direction.

A flow control valve restricts the flow rate at which pressurised fluid is transferred in a hydraulic circuit.

A pressure control valve limits or reduces the pressure level of the fluid or generates a control signal when a set pressure in some system part has been reached for initiating a subsequent action. The types of pressure control valves include pressure reducing valves, unloading valves, sequence valves, counterbalance valves, and brake valves.

A directional control valve is specified as a 'port/position valve', where the 'port' represents the number of ports, and the 'position' represents the number of switching positions of the valve. The ports of hydraulic valves are designated using a letter system following the ISO 4401 standard.

3/2-DC Valve

Figure 5.1 shows the cross-sectional views of a 3/2-directional (DC) valve [that is, 3 ports and 2 switching positions] in the normal and actuated positions. In the normal position of the valve, the working port A is closed to the pressure port P, and open to the tank port T. The pressure port P is blocked in the normal position of the valve. In the actuated position of the valve, the working port A is open to the pressure port P and closed to the tank port T. The 3/2-way valves can be used to control single-acting hydraulic cylinders and other valves.

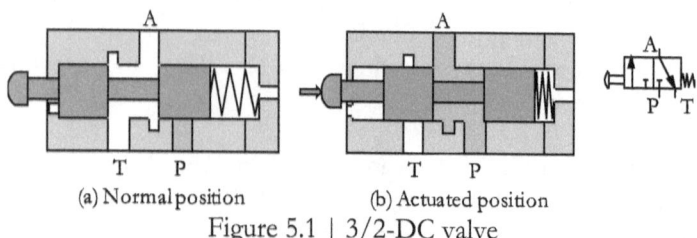

(a) Normal position   (b) Actuated position

Figure 5.1 | 3/2-DC valve

4/2-DC Valve

Figure 5.2 shows the cross-sectional views of a 4/2-directional (DC) valve in the normal and actuated positions. In the normal position of the valve, paths from the port P to the working port B and from the working port A to the port T are open. When the valve is actuated, paths from the port P to the working port A and from the working port B to the port T are open. This valve can be used as the final control element to drive a double-acting hydraulic cylinder or bi-directional hydraulic motor.

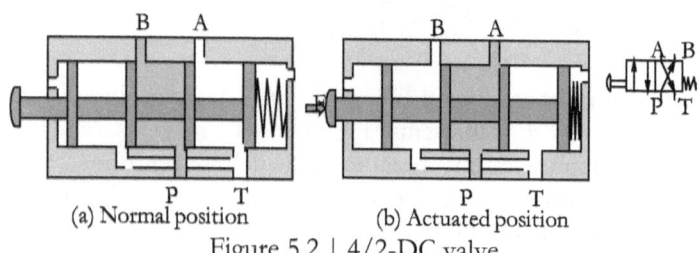

(a) Normal position   (b) Actuated position

Figure 5.2 | 4/2-DC valve

4/3-DC valves

A 4/3-DC (way) valve has four ports and three switching positions. In the normal position, the valve remains in the centre position using springs mounted at either end of its spool shaft. The centre position is the position the valve assumes when there is no movement of the associated hydraulic component required.

Centre Positions of 4/3-DC Valves

Apart from the two usual switching positions of the valve for the forward and return strokes of the actuator, an additional spring-centred switching position can be incorporated in the valve to realise certain additional requirements. These additional requirements have resulted in the development of 4/3-DC valves. The symbols of various centre positions of 4/3-DC valves are given in Figure 5.3.

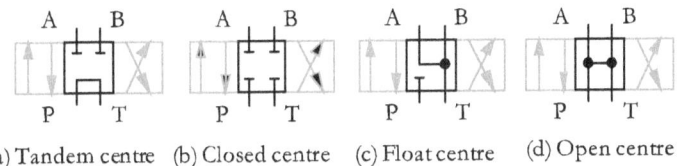

(a) Tandem centre   (b) Closed centre   (c) Float centre   (d) Open centre

Figure 5.3 | Various centre positions of 4/3-DC valves

Check Valve

A check valve permits the flow readily in one direction and prevents the flow entirely in the reverse direction. The valve, as shown in Figure 5.4, consists of a valve body and a spring-biased ball poppet or cone poppet, apart from inlet and outlet ports ('A' and 'B'). The spring holds the poppet against the valve seat.

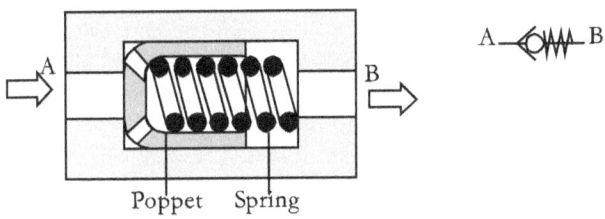

Figure 5.4 | Two designs of check valves

Pilot-operated check valve

A pilot-operated check valve, as shown in Figure 5.5, is a variant of the primary check valve. It consists of a valve body, a poppet biased by a spring, ports 'A' and 'B', and a pilot port 'X'. The poppet has a pilot piston attached to the poppet stem. Pressure applied to the port A causes the poppet to lift from its seat against the spring force, allowing the fluid flow through the valve from the port A to the port B with a low-pressure drop across it.

The flow through the valve is usually blocked when the intended flow direction is from port B to the port A, by poppet reseating. However, the poppet can be held open by the application of an external pilot signal through the pilot port X. Considering the forces developed by the pressures and areas on both sides of the poppet, the pressure applied to the pilot port displaces the poppet from its seat, allowing fluid flow from the port B to the port A.

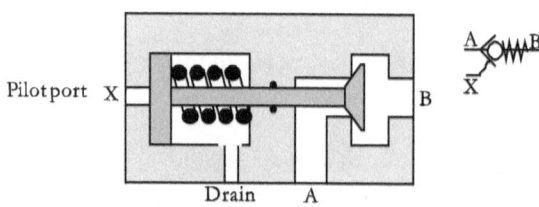

Figure 5.5 | A pilot-operated check valve

Flow Control Valves

An essential function of the flow control valve is to offer a hydraulic resistance to the flow of the system fluid and hence to control the volume of the fluid passing a given point in the system. The flow control function in both directions can be realised using a controlled restriction in the path of the fluid flow (Figure 5.6).

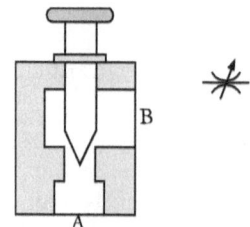

Figure 5.6 | A pilot-operated check valve

One-way Flow Control Valve

The basic flow control valve, such as the elementary throttle valves, as discussed in the previous section, provide the same flow restriction in either direction of the fluid flow. A one-way flow control valve provides a restricted flow of the fluid in one direction and an unrestricted flow in the reverse direction.

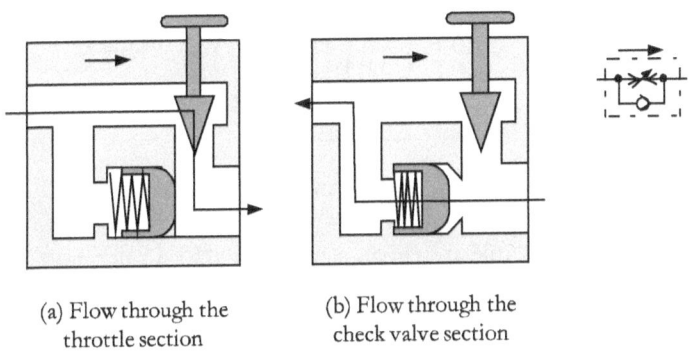

(a) Flow through the throttle section

(b) Flow through the check valve section

Figure 5.7 | Two positions of a one-way flow control valve

The one-way flow control valve consists of a throttling section and a check valve section, both integrated into its valve body in a parallel arrangement. It is also known as a throttle check valve. The annular gap at the throttling point can be controlled by turning the throttling screw. Figure 5.7 gives the schematic diagrams representing the two positions of the one-way flow control valve.

The check valve blocks the system flow in one direction forcing the fluid to flow through the controlled cross-section, as shown in Figure 5.7(a). Hence, the flow is throttled, and the flow rate is controlled, in that direction. In the opposite direction, the fluid flows freely through the opened check valve without any restriction, as shown in Figure 5.7(b). This valve is used, when a direction-sensitive speed control of a hydraulic actuator is required as in clamping applications.

Unloading Valves

An unloading valve (Figure 5.8) is a type of pressure relief valve that can be used to unload a pump delivery when the critical pressure at the downstream side has been reached.

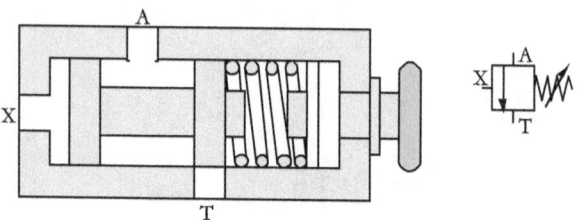

Figure 5.8 | An unloading valve

It consists of an inlet port 'A', a tank port 'T', a pilot port 'X', and a spring-biased spool. In the normal position, the valve remains closed by the spring force. The unloading pressure can be varied by adjusting the spring tension. The pilot port is provided to accept the external pressure signal, which acts on one end of the spring-biased valve spool.

The spool is shifted, and the passage from the port A to the tank port T is opened when a signal is applied to the pilot port X. Therefore, the pump delivery can be diverted to the tank through this passage at a low-pressure. The spool is kept open until the pilot pressure goes below the spring bias. The use of this valve minimizes the heat generation that leads to the cost-effective operation of the system.

Applications of Unloading Valves: Unloading valves are designed for use in accumulator circuits, hydraulic motor circuits, and two-pump 'hi-lo' circuits. For example, a hydraulic circuit with an accumulator can also be unloaded, when the accumulator is charged to a specified level, using an unloading valve.

Chapter 6 | Hydraulic Filters

Filters (Figure 6.1) are devices for removing particulate contamination from hydraulic systems. A power pack is usually equipped with many filter units. A filter can be equipped with a bypass valve for protection against filter burst. A contamination indicator can be incorporated to get an indication when the filter is fully clogged. According to their locations in a hydraulic system, filters are categorised into: (1) Suction strainer/filter, (2) Pressure filter, (3) Return-line filter, and (4) Off-line filter.

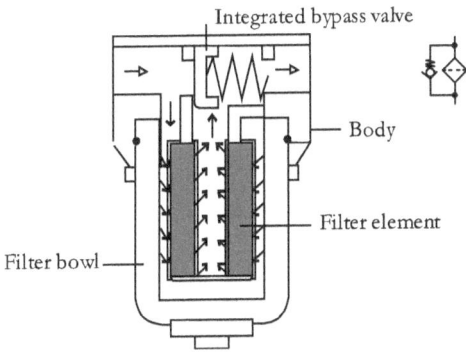

Figure 6.1 | Hydraulic Filters

Beta Ratio
The Beta (ß) ratio signifies the effectiveness of the filter element in removing the contaminants from the associated hydraulic system. It can then be determined by using the formula given below:

$$\text{Beta ratio}_{x(c)} = \frac{\text{Particle count in the upstream fluid}}{\text{Particle count in the downstream fluid}}$$

Where the subscript 'x' stands for the specified particle size and subscript '(c)' refers to the 'certified' calibration. That is, the subscript, (c), indicates the test adherence to the new ISO 16889 standard.

Filter Efficiency

The filter efficiency represents the ability of the filter to remove contaminants of a specified size under the specified test conditions. This efficiency, in terms of percentage, can be found by the following equation:

$$\text{Efficiency}_{(x)} = \left(1 - \frac{1}{\text{ß}}\right) \times 100$$

Absolute Micron Rating of a filter is the smallest size of particles it can capture in excess of 98.6% on the first pass through it.

Nominal Micron Rating of a filter is the smallest micron size of particles it can capture in a specified quantity, in the range from 50% to 95% on the first pass through it.

Typical specifications of filters are presented in Table 6.1:

Table 6.1 | Filter parameters

Parameter	Choice
Filter element media	Stainless wire mesh, Stainless fibre, Cellulose, Inorganic glass fibre
Micron rating	3, 5, 10, 20, 25, 50, 100, 200 µm
Flow rate	60, 110, 160, 240, 330, 475, 660, 990 lpm
Maximum operating pressure	Depends on the type of filters, namely suction filters, pressure filters, and return-line pressures
Seal material	NBR, FPM (Viton), EPDM (Ethylene Propelene Diene Monomer)
Valve	By-pass valve
By-pass setting	0.1 to 7 bar
Clogging indicator	Visual, Electrical, Pressure gauge
Port size	NPT ¾ to 3, G ½ to 2, GAS and SAE Flanges, SAE O-ring

Chapter 7 | Hydraulic Accumulators

An accumulator is a device used for absorbing shock pressures and storing energy in a hydraulic system. It mainly consists of a vessel in which a hydraulic fluid is held under pressure by a raised weight or a spring or a volume of compressed gas. It is, thus, possible to store potential energy in the accumulator, when the associated system pressure remains greater than that of the accumulator. The accumulator can release the stored energy back into the system for performing some useful hydraulic task when the system pressure falls below that of the accumulator. A power pack may be equipped with accumulators to cover peak requirements. All accumulators must comply with work safety regulations. Figure 7.1 shows the typical accumulators.

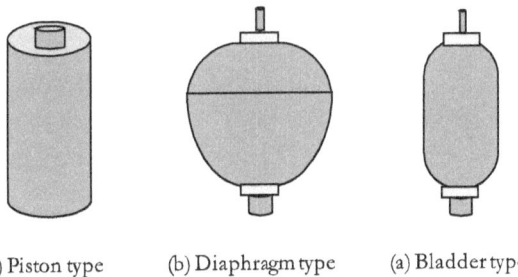

(a) Piston type   (b) Diaphragm type   (a) Bladder type

Figure 7.1 | Accumulators

Piston Accumulators
A piston accumulator [Figure 7.1(a)] consists of a cylinder (or shell) with a finely finished internal surface and a freely floating and light-weight metal (typically aluminium) piston. The movable piston effectively separates the cylinder into a fluid section and a gas section. The gas section is pre-charged with dry nitrogen gas. The fluid part is connected to the hydraulic system so that the piston accumulator draws the fluid when the system pressure goes up, thus compressing the gas. When the system pressure drops, the pressurised gas expands and forces the fluid back into the system.

Diaphragm Accumulators

A diaphragm (membrane) accumulator [Figure 7.1(b)] consists of two steel hemispheres (shells) and a flexible synthetic diaphragm secured to the shells. It also consists of a gas chamber and a fluid chamber. The diaphragm is used to separate the gas and the fluid sections. The gas chamber is pre-charged to a definite pressure, and the fluid chamber is connected to the hydraulic system.

Bladder Accumulators

A bladder (or bag) accumulator [Figure 7.1(c)] consists of a seamless cylindrical pressure vessel (shell) and an internal elastomeric bladder (bag). The bladder divides the shell into two chambers, namely, the fluid chamber on the system side and the gas chamber inside the bladder. The gas chamber is pre-charged to a certain pressure level.

Pre-charging of Accumulators

In general, a gas accumulator is pre-charged to a certain percentage of the minimum system pressure, depending upon the type of accumulator and application, and as per the recommendation of its manufacturer. Typically, the pre-charge pressure is 80 to 90 per cent of the minimum working (operating) pressure of the system for energy storage applications. The pre-charge pressure for a pulsation compensator or a shock absorber can typically be taken as 65 to 80 per cent of the minimum operating pressure.

Control and Safety Block

Safety devices must be incorporated in an accumulator to provide shut-off facility and pressure-limiting and pressure relief features. So, it is recommended to use a safety-and-shut-off block with an accumulator for protecting the system and the personnel against the hazardous stored energy.

An accumulator is a pressure vessel that stores an enormous amount of potential energy. Accumulators can be dangerous to personnel and property if they discharge stored pressure inadvertently. It is necessary to isolate the accumulators from the Pump and discharge the trapped pressure, during maintenance/emergency period. Safety devices (Figure 7.2) must be incorporated to provide a shut-off facility and bleed-off facility and pressure relief features. Accumulators are subject to regulations applicable to the one's region.

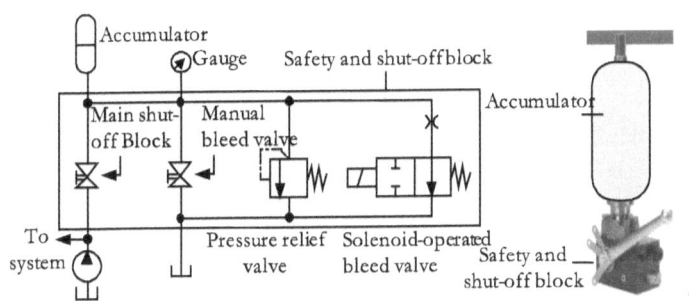

Figure 7.2 | Accumulator with control and safety block

Accumulators can be installed vertically, horizontally, or at any angle depending upon the application requirements. However, in a system where contamination is a severe problem, a vertical mount with the fluid port of the accumulator oriented downward is, usually, preferred.

The specifications for the safety block is given in Table 7.1:

Table 7.1 | The specifications for the safety block

Pressure rating	207 bar / 340 bar
Safety shut-off	Manual
Bleed valve	Manual / Solenoid
Solenoid control voltage	24 V DC / 115 AC, 60 Hz
Seal material	Buna – N, Viton

Table 7.2 gives a comparison of diaphragm, bladder and piston type hydraulic accumulators.

Table 7.2 | Comparison of accumulators

Parameter / property	Diaphragm	Bladder	Piston
Size	Up to 3.5 l	Up to 54 l	Up to 1350 l
Working pressure	250 bar	690 bar	2500 bar
Flow rate	Up to 150 lpm	Up to 900 lpm	Up to 900 lpm
Seal materials	Buna-N Butyl	Buna-N Butyl Viton	Buna-N Viton
Temperature limits (Typical)	-40°C to 176°C	-29°C to 93°C	-29°C to 74°C
Heaviness	Light-weight	Medium-weight	Heavy
Cost	Low	Medium	High
Application	Suitable for small volume and flow rates	Best for general purpose applications	Best for large volumes or high flow rates
Shock suppression ability	Good shock absorber	Good shock absorber	Not suitable for suppressing shocks
Mounting position	Any position	Any position	Only vertical position
Oil port connection	SAE NPT	SAE NPT	SAE NPT

Chapter 8 | Power Pack Control Unit

A power pack may contain multiple pumps (operating pump/standby pump), accumulators, a cooling fan, heater, etc. A control box with a controller, power switches, and display elements is necessary for the control of the power pack.

The control system (Figure 8.1) may cover the following functions: (1) Starting the standby pump, (2) Changing the operating pump, (3) Switching the accumulator charging valve, (4) Starting the cooling fan, (5) Stopping the heater, and (6) Generating alarm and message signals.

In case of excessive temperature rise of the fluid, the pump-motor unit is switched off. The heating system, if any, will operate depending on the temperature of the hydraulic fluid.

If the pump fails to charge the accumulator, the standby pump starts automatically and switches off as soon as the pressure is normal.

If the accumulator pressure drops too low, an alarm indicates hydraulic power unit failure.

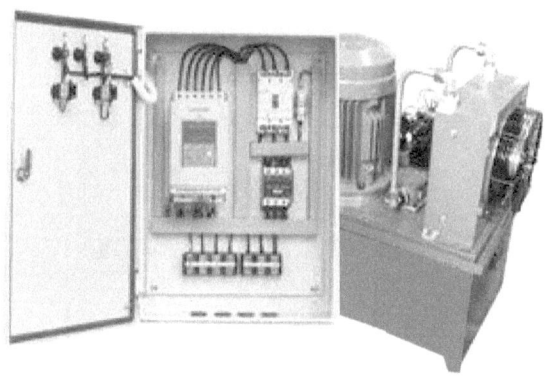

Figure 8.1 | Power pack with a control unit

Chapter 9 | Noise in Hydraulic Systems

Industries are increasingly becoming alarmed about subjecting their workforce to plant noise. Therefore, the need for a quiet work environment in industries is growing. Like any other systems, hydraulic systems may be a source of the high noise.

The intensity of the sound is measured in terms of 'decibel' (dB). The frequency is measured in hertz (Hz). The frequency of the audible sound for human ears is in the range between 20 Hz to 20,000 Hz. Human beings are very much sensitive to the sound in the frequency range from 1 kHz to 4 kHz, rather than to low-frequency or high-frequency sounds. For this reason, the sound meter is usually fitted with a special filter, such as 'A-weighting' filter, whose response to frequency is similar to that of the human ear. If an A-weighting filter is used, the sound pressure level is given in dB (A).

Figure 9.1 gives a typical self-explanatory noise level characteristic of a hydraulic pump. In general, noise is generated in the workplace by the machines, blast furnaces, power saws, ventilation systems, circulation pumps, and vehicle engines.

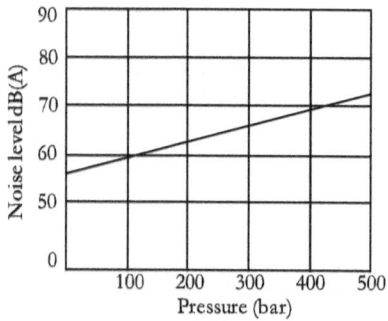

Figure 9.1 | A typical characteristic of the noise level Vs pressure characteristic of a hydraulic pump

Sources of Noise in Hydraulic Systems

There are many sources of noise in hydraulic systems. In general, higher noise levels in hydraulic systems are produced by the piston pumps and the relief valves. The sources of noise are categorised as: (1) structure-borne noise, (2) fluid-borne noise, and (3) airborne noise.

The structure-borne noise in a hydraulic system is produced mainly by the vibrating mass of the hydraulic pump, electric motor, couplings, and belt drives. The pump noise usually originates from cavitation, air present in the system fluid, and the rolling and sliding action of bearings and damaged pump elements. Further, the electric motor noise stems from its stator and rotor assembly, bearings, and fan.

Another cause for the noise is the misalignment of the coupling between the pump and motor. Next, any belt drive in the system can also give off low-frequency noises. The loosened bolts in a machine may be the reason for its increased vibration. The worn-out poppet or the seat in a pressure relief valve can also cause excessive noise.

Another factor that can contribute much to the noise level is the improper use of fluid conductors, such as hoses. The fluid conductors can provide the primary path for the propagation of the noise from the system pump to various other system components.

The positive-displacement pumps in the system produce pulsating fluid delivery resulting in a corresponding sequence of turbulence and pressure pulsations. These pulsations can create the fluid-borne noise, which causes the downstream components of the system to vibrate. The transfer of the structure-induced and fluid-induced vibrations to the adjacent air medium sets off the air-borne noise. Moreover, the vibrations in solids and liquids can travel a considerable distance before producing the airborne noise.

Noise Reduction Techniques

It is necessary to design hydraulic systems, especially the power units with appropriate noise reduction techniques to reduce the damaging effects of noise. In general, the noise in the machine can be controlled by: (1) using quieter work processes, (2) enclosing the machine completely to reduce the noise at source, and (3) using sound-absorbing materials in the machine to prevent the spread of the noise. The noise reduction can be achieved by isolating the motor-pump unit of the machine from its base, isolating the structural elements of the power unit that could intensify the sound, and using the hoses and tubing correctly. An integrated motor-pump unit has very low sound levels. Many other factors, such as the mounting, tank style, and plant layout, affect the noise levels. The following activities are recommended to reduce the noise and vibration in a hydraulic machine/system.

- Select proper frame and base,
- Enclose noisy machine parts,
- Use resilient mounts for the machine, whenever possible,
- Use densely perforated plates in the machine,
- Select pumps and drive motors with low noise level ratings,
- Align the pump-motor unit in the system properly,
- Overhaul/replace the worn or damaged pump unit in the system,
- Replace any broad belt in the system with a set of narrow belts, separated by spacers,
- Correctly route the hoses,
- Avoid long, and unsupported conductor runs,
- Reduce cavitation,
- Prevent entry of air into the system fluid,
- Check the conditions of seals, bearings, and couplings of the system components for wear or misalignment, and
- Overhaul/replace the worn-out poppet or seals in the control valves.

Chapter 10 | Power Packs with Standardized Sub-assemblies

Power packs can be configured and built with standardised sub-assemblies or blocks. The basic block is the oil tank with cover. Components like a pump, electric motor, etc. can be fitted to the oil tank. The base block with pressure relief valve can be attached to the pump output. Many other standard blocks with pressure relief valve, check valve, return-line filter, directional control valve, throttle-check valve, pilot-operated check valve, unloading valve, etc. are available for the modular construction of various hydraulic circuits in the form of vertical and/or horizontal stacking assembly. The power unit can also be equipped with accumulators. The required combination of components can be selected based on the tank capacity, pump type, flow rate, pressure, size of the electric motor, type of pressure control etc. Figure 10.1 shows the schematic of a power pack with standardised sub-assemblies and the relevant control circuit.

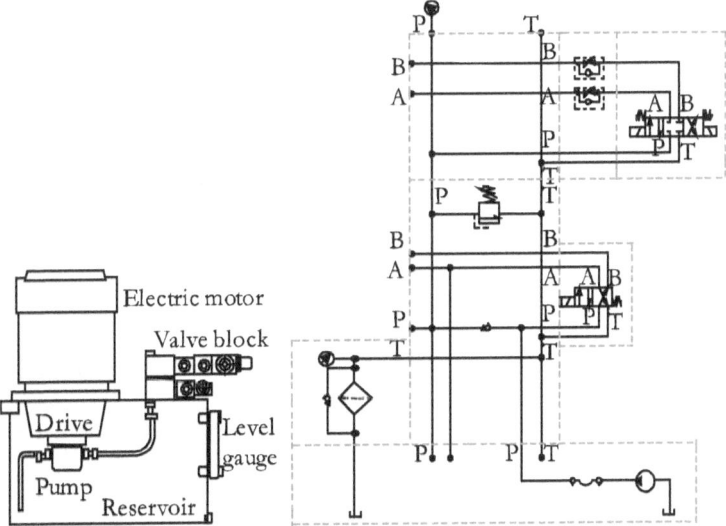

Figure 10.1 | A power pack with standardised sub-assemblies and the relevant control circuit

11 | Review Questions

1. What is a hydraulic power pack? Describe the different parts used in power packs with their functions.

2. Draw the simple sketch of a hydraulic reservoir, and explain its main parts.

3. What are the essential and optional components of hydraulic reservoirs?

4. What are the features included in the power pack of a hydraulic system to control the contamination and the system temperature?

5. Give five factors do you consider in the design of hydraulic reservoirs.

6. What is the customary rule for finding the size of a hydraulic reservoir?

7. What are the circumstances under which the size of the reservoir for a hydraulic system is larger than usual?

8. Briefly explain the various pump-tank layouts in hydraulic systems.

9. What are the different types of heat exchangers used in hydraulic systems? Explain briefly.

10. How is the power developed in a pump related to the pump flow rate and the system pressure?

11. Briefly explain the following hydraulic terms: (1) Pump displacement and (2) Pump delivery.

12. Classify positive-displacement hydraulic pumps with suitable examples.

13. Differentiate between fixed-displacement and variable-displacement hydraulic pumps.

14. Explain the essential constructional features of hydraulic gear/vane/piston pumps.

15. Explain some critical performance measures for hydraulic pumps.

16. Describe the classification and performance features of different types of hydraulic pumps.

17. Draw the typical characteristic curves of hydraulic pumps and explain.

18. Why is it essential to have a relief valve in a hydraulic system with a positive displacement pump?
19. Describe the operation of a basic PRV used in a hydraulic system with the help of a suitable circuit.
20. What are the consequences of malfunction in a pressure relief valve in a hydraulic system?
21. List out some techniques for reducing the noise produced by hydraulic power sources.

12 | Numerical Problems

1. Calculate the amount of surface area required for a hydraulic reservoir to obtain a heat dissipation capacity of 0.088 kW, if the ambient temperature is 20 ^{0}C and the maximum desired temperature is 60 ^{0}C.
2. What is the displacement of a hydraulic pump required to produce 0.000505 m^3/s while turning at a speed of 1750 rpm? [Ans: 0.0000173 m^3/rev]
3. A hydraulic pump is driven by a 5 kW motor. The pump operates at a pressure of 100 bar. Assuming no losses, what is the flow rate through the system in units of m^3/s? [Ans: kWx1000/Pa = 0.0005 m^3/s]
4. A fixed-displacement pump, driven at a speed of 1200 rpm, is supplied with an actual torque of 23000 Nm. What is the mechanical efficiency of the pump operating with a theoretical flow rate of 0.2 m^3/s at 170 bar? Also calculate the volumetric and overall efficiency of the pump, if the pump slippage is 0.005 m^3/min.
[Ans: η_m=85%, η_v=97.5%, η_o= 82.22%]

Appendix 1

Pump Data

Appendix 1 gives the typical pump data extracted from the manufacturer's domain.

Pump Data – External Gear Pumps

Table A1.1 | External gear pumps

Motor kW	Pump type	Speed (rpm)	Displacement (cc/rev)	Flow rate (lpm)	Pressure (Cont) (bar)
0.55	Ext Gear	1395	0.8	1.1	200
0.55	Ext Gear	1395	1.2	1.6	170
0.55	Ext Gear	1395	1.6	2.1	125
0.55	Ext Gear	1395	2.1	2.8	95
0.55	Ext Gear	1395	2.5	3.3	80
0.55	Ext Gear	1395	3.3	4.4	60
0.55	Ext Gear	1395	3.6	4.8	55
0.55	Ext Gear	1395	4.4	5.8	45
0.55	Ext Gear	1395	4.8	6.4	40
0.55	Ext Gear	1395	5.8	7.7	35
0.55	Ext Gear	1395	6.2	8.2	30
0.55	Ext Gear	1395	7.9	10.5	25
0.55	Ext Gear	2790	0.8	2.1	125
0.55	Ext Gear	2790	1.2	3.1	85
0.55	Ext Gear	2790	1.6	4.2	60
0.55	Ext Gear	2790	2.1	5.6	45
0.55	Ext Gear	2790	2.5	6.6	40
0.55	Ext Gear	2790	3.3	8.7	30
0.55	Ext Gear	2790	3.6	9.5	30
0.55	Ext Gear	2790	4.4	11.7	25
0.55	Ext Gear	2790	4.8	12.7	20
0.55	Ext Gear	2790	5.8	15.4	20
0.55	Ext Gear	2790	6.2	16.4	15
0.55	Ext Gear	2790	7.9	20.9	15

Pump Data – External Gear Pumps

Table A1.2 | External gear pumps

Motor kW	Pump type	Speed (rpm)	Displacement (cc/rev)	Flow rate (lpm)	Pressure (Cont) (bar)
0.75	Ext Gear	1395	1.2	1.6	200
0.75	Ext Gear	1395	1.6	2.1	170
0.75	Ext Gear	1395	2.1	2.8	130
0.75	Ext Gear	1395	2.5	3.3	110
0.75	Ext Gear	1395	3.3	4.4	80
0.75	Ext Gear	1395	3.6	4.8	75
0.75	Ext Gear	1395	4.4	5.8	60
0.75	Ext Gear	1395	4.8	6.4	55
0.75	Ext Gear	1395	5.8	7.7	45
0.75	Ext Gear	1395	6.2	8.2	45
0.75	Ext Gear	1395	7.9	10.5	35
0.75	Ext Gear	2850	0.8	2.2	165
0.75	Ext Gear	2850	1.2	3.2	110
0.75	Ext Gear	2850	1.6	4.3	85
0.75	Ext Gear	2850	2.1	5.7	65
0.75	Ext Gear	2850	2.5	6.8	55
0.75	Ext Gear	2850	3.3	8.9	40
0.75	Ext Gear	2850	3.6	9.7	35
0.75	Ext Gear	2850	4.4	11.9	30
0.75	Ext Gear	2850	4.8	13.0	30
0.75	Ext Gear	2850	5.8	15.7	25
0.75	Ext Gear	2850	6.2	16.8	20
0.75	Ext Gear	2850	7.9	21.4	15

Pump Data – External Gear Pumps

Table A1.3 | External gear pumps

Motor kW	Pump type	Speed (rpm)	Displacement (cc/rev)	Flow rate (lpm)	Pressure (Cont) (bar)
1.1	Ext Gear	1410	1.6	2.1	200
1.1	Ext Gear	1410	2.1	2.8	190
1.1	Ext Gear	1410	2.5	3.3	160
1.1	Ext Gear	1410	3.3	4.4	120
1.1	Ext Gear	1410	3.6	4.8	110
1.1	Ext Gear	1410	4.0	5.4	100
1.1	Ext Gear	1410	4.4	5.9	90
1.1	Ext Gear	1410	4.8	6.4	80
1.1	Ext Gear	1410	5.0	6.7	80
1.1	Ext Gear	1410	5.8	7.8	70
1.1	Ext Gear	1410	6.2	8.3	65
1.1	Ext Gear	1410	6.3	8.5	65
1.1	Ext Gear	1410	7.9	10.6	50
1.1	Ext Gear	1410	8.0	10.8	50
1.1	Ext Gear	1410	10.0	13.5	40
1.1	Ext Gear	1410	12.5	16.9	30
1.1	Ext Gear	1410	16.0	21.6	25
1.1	Ext Gear	1410	20.0	27.0	20
1.1	Ext Gear	1410	25.0	33.7	15
1.1	Ext Gear	2835	0.8	2.2	200
1.1	Ext Gear	2835	1.2	3.2	160
1.1	Ext Gear	2835	1.6	4.3	125
1.1	Ext Gear	2835	2.1	5.7	95
1.1	Ext Gear	2835	2.5	6.7	80
1.1	Ext Gear	2835	3.3	8.9	60
1.1	Ext Gear	2835	3.6	9.7	55
1.1	Ext Gear	2835	4.4	11.9	45
1.1	Ext Gear	2835	4.8	12.9	40
1.1	Ext Gear	2835	5.8	15.6	35
1.1	Ext Gear	2835	6.2	16.7	30
1.1	Ext Gear	2835	7.9	21.3	25

Pump Data – External Gear Pumps

Table A1.4 | External gear pumps

Motor kW	Pump type	Speed (rpm)	Displacement (cc/rev)	Flow rate (lpm)	Pressure (Cont) (bar)
1.5	Ext Gear	1410	2.1	2.8	200
1.5	Ext Gear	1410	2.5	3.3	200
1.5	Ext Gear	1410	3.3	4.4	165
1.5	Ext Gear	1410	3.6	4.8	150
1.5	Ext Gear	1410	4.0	5.4	135
1.5	Ext Gear	1410	4.4	5.9	120
1.5	Ext Gear	1410	4.8	6.4	110
1.5	Ext Gear	1410	5.0	6.7	110
1.5	Ext Gear	1410	5.8	7.8	95
1.5	Ext Gear	1410	6.2	8.3	85
1.5	Ext Gear	1410	6.3	8.5	85
1.5	Ext Gear	1410	8	10.8	65
1.5	Ext Gear	1410	10	13.5	55
1.5	Ext Gear	1410	12.5	16.9	45
1.5	Ext Gear	1410	16.0	21.6	35
1.5	Ext Gear	1410	20.0	27.0	25
1.5	Ext Gear	1410	25.0	33.7	20
1.5	Ext Gear	1410	7.9	10.6	70
1.5	Ext Gear	2860	1.2	3.2	200
1.5	Ext Gear	2860	1.6	4.3	165
1.5	Ext Gear	2860	2.1	5.7	125
1.5	Ext Gear	2860	2.5	6.8	105
1.5	Ext Gear	2860	3.3	9.0	80
1.5	Ext Gear	2860	3.6	9.7	75
1.5	Ext Gear	2860	4.4	11.9	60
1.5	Ext Gear	2860	4.8	13.0	55
1.5	Ext Gear	2860	5.8	15.8	45
1.5	Ext Gear	2860	6.2	16.8	45
1.5	Ext Gear	2860	7.9	21.5	35

Pump Data – External Gear Pumps

Table A1.5 | External gear pumps

Motor kW	Pump type	Speed (rpm)	Displacement (cc/rev)	Flow rate (lpm)	Pressure (Cont) (bar)
2.2	Ext Gear	1420	3.3	4.4	200
2.2	Ext Gear	1420	3.6	4.8	200
2.2	Ext Gear	1420	4.0	5.4	195
2.2	Ext Gear	1420	4.4	5.9	180
2.2	Ext Gear	1420	4.8	6.5	165
2.2	Ext Gear	1420	5.0	6.7	155
2.2	Ext Gear	1420	5.8	7.8	135
2.2	Ext Gear	1420	6.2	8.4	125
2.2	Ext Gear	1420	6.3	8.5	125
2.2	Ext Gear	1420	7.9	10.7	100
2.2	Ext Gear	1420	8.0	10.8	100
2.2	Ext Gear	1420	10.0	13.5	80
2.2	Ext Gear	1420	12.5	16.9	65
2.2	Ext Gear	1420	16.0	21.6	50
2.2	Ext Gear	1420	20.0	27.0	40
2.2	Ext Gear	1420	25.0	33.7	30
2.2	Ext Gear	2850	1.6	4.3	200
2.2	Ext Gear	2850	2.1	5.7	185
2.2	Ext Gear	2850	2.5	6.8	155
2.2	Ext Gear	2850	3.3	8.9	120
2.2	Ext Gear	2850	3.6	9.7	110
2.2	Ext Gear	2850	4.4	11.9	90
2.2	Ext Gear	2850	4.8	13.0	80
2.2	Ext Gear	2850	5.8	15.7	65
2.2	Ext Gear	2850	6.2	16.8	65
2.2	Ext Gear	2850	7.9	21.4	50

Pump Data – External Gear Pumps

Table A1.6 | External gear pumps

Motor kW	Pump type	Speed (rpm)	Displacement (cc/rev)	Flow rate (lpm)	Pressure (Cont) (bar)
3.0	Ext Gear	1420	4.0	5.4	270
3.0	Ext Gear	1420	4.4	5.9	200
3.0	Ext Gear	1420	4.8	6.5	200
3.0	Ext Gear	1420	5.0	6.7	215
3.0	Ext Gear	1420	5.8	7.8	160
3.0	Ext Gear	1420	6.2	8.4	160
3.0	Ext Gear	1420	6.3	8.5	170
3.0	Ext Gear	1420	7.9	10.7	135
3.0	Ext Gear	1420	8.0	10.8	135
3.0	Ext Gear	1420	10.0	13.5	105
3.0	Ext Gear	1420	12.5	16.9	85
3.0	Ext Gear	1420	16.0	21.6	65
3.0	Ext Gear	1420	17.0	22.9	65
3.0	Ext Gear	1420	20.0	27.5	55
3.0	Ext Gear	1420	25.0	33.7	45
3.0	Ext Gear	1420	27.0	36.9	40
3.0	Ext Gear	1420	34.0	45.9	30
3.0	Ext Gear	2895	3.3	9.1	160
3.0	Ext Gear	2895	3.6	9.9	145
3.0	Ext Gear	2895	4.4	12.0	120
3.0	Ext Gear	2895	4.8	13.2	110
3.0	Ext Gear	2895	5.8	16.0	90
3.0	Ext Gear	2895	6.2	17.1	85
3.0	Ext Gear	2895	7.9	21.7	65

Pump Data – External Gear Pumps

Table A1.7 | External gear pumps

Motor kW	Pump type	Speed (rpm)	Displacement (cc/rev)	Flow rate (lpm)	Pressure (Cont) (bar)
4.0	Ext Gear	1455	5.0	6.8	270
4.0	Ext Gear	1455	6.3	8.6	225
4.0	Ext Gear	1455	8.0	11.0	175
4.0	Ext Gear	1455	10.0	13.8	140
4.0	Ext Gear	1455	12.5	17.3	110
4.0	Ext Gear	1455	16.0	22.1	90
4.0	Ext Gear	1455	17.0	23.3	85
4.0	Ext Gear	1455	20.0	27.5	70
4.0	Ext Gear	1455	25.0	34.6	55
4.0	Ext Gear	1455	27.0	36.9	50
4.0	Ext Gear	1455	34.0	46.5	40

Table A1.8 | External gear pumps

Motor kW	Pump type	Speed (rpm)	Displacement (cc/rev)	Flow rate (lpm)	Pressure (Cont) (bar)
5.5	Ext Gear	1455	8.0	11.0	240
5.5	Ext Gear	1455	10.0	13.8	190
5.5	Ext Gear	1455	12.5	17.3	155
5.5	Ext Gear	1455	16.0	22.1	120
5.5	Ext Gear	1455	17.0	23.5	110
5.5	Ext Gear	1455	20.0	27.5	95
5.5	Ext Gear	1455	25.0	34.6	75
5.5	Ext Gear	1455	27.0	37.3	70
5.5	Ext Gear	1455	34.0	47.0	55

Pump Data – External Gear Pumps

Table A1.9 | External gear pumps

Motor kW	Pump type	Speed (rpm)	Displacement (cc/rev)	Flow rate (lpm)	Pressure (Cont) (bar)
7.5	Ext Gear	1455	10.0	13.8	250
7.5	Ext Gear	1455	12.5	17.3	210
7.5	Ext Gear	1455	16.0	22.1	165
7.5	Ext Gear	1455	17.0	23.5	155
7.5	Ext Gear	1455	20.0	27.5	130
7.5	Ext Gear	1455	25.0	34.6	105
7.5	Ext Gear	1455	27.0	37.3	95
7.5	Ext Gear	1455	34.0	47.0	75

Table A1.10 | External gear pumps

Motor kW	Pump type	Speed (rpm)	Displacement (cc/rev)	Flow rate (lpm)	Pressure (Cont) (bar)
11.6	Ext. Gear	4000	5.3	18.7	276
14.3	Ext. Gear	4000	6.5	22.9	276
18.2	Ext. Gear	4000	8.3	29.2	276
20.4	Ext. Gear	3600	10.3	32.6	276
23.4	Ext. Gear	3300	12.9	37.5	276
26.5	Ext. Gear	3000	16.1	44.4	276
27.5	Ext. Gear	2800	20.0	51.5	250
28.8	Ext. Gear	2600	24.0	57.4	235
25.6	Ext. Gear	2300	28.4	60.1	200
23.4	Ext. Gear	2100	33.4	64.5	170

Pump Data – Internal Gear Pumps

Table A1.11 | Internal gear pumps
(Pressure Range – 100 bar | 125 bar | 160 bar)

Motor kW	Pump type	Speed (rpm)	Displacement (cc/rev)	Flow rate (lpm)	Pressure (Cont) (bar)
4.1	Int. Gear	4500	10.3	14.9	160
4.1	Int. Gear	4000	12.6	18.3	125
4.1	Int. Gear	3600	15.9	23.0	100
8.2	Int. Gear	3600	20.0	29.0	160
8.2	Int. Gear	3250	25.3	36.7	125
8.2	Int. Gear	3000	31.2	45.2	100
16.4	Int. Gear	3000	40.7	59.0	160
16.4	Int. Gear	2600	50.3	72.9	125
16.4	Int. Gear	2300	64.7	93.6	100
32.8	Int. Gear	2300	78.6	114	160
32.8	Int. Gear	2100	101.1	146	125
32.8	Int. Gear	1800	127.3	184	100
64.6	Int. Gear	1800	160.5	232	160
64.6	Int. Gear	1800	202.1	293	125
64.6	Int. Gear	1800	249.7	362	100
129.1	Int. Gear	1800	326.0	472	160
129.1	Int. Gear	1800	402.6	583	125
129.1	Int. Gear	1500	498.5	722	100

Pump Data – Internal Gear Pumps

Table A1.12 | Internal gear pumps (Pressure Range - 210 bar)

Motor kW	Pump type	Speed (rpm)	Displacement (cc/rev)	Flow rate (lpm)	Pressure (Cont) (bar)
2.7	Int. Gear	5000	5.1	7.4	210
3.2	Int. Gear	5000	6.3	9.1	210
4.3	Int. Gear	5000	8.0	11.5	210
5.4	Int. Gear	4300	10.0	14.5	210
6.5	Int. Gear	4300	12.6	18.3	210
8.6	Int. Gear	4300	15.6	22.6	210
10.8	Int. Gear	3600	20.4	29.5	210
13.4	Int. Gear	3600	25.1	36.4	210
17.2	Int. Gear	3600	32.4	46.8	210
21.5	Int. Gear	3000	39.3	56.9	210
26.9	Int. Gear	3000	50.6	73.2	210
33.9	Int. Gear	3000	63.7	92.1	210
43.0	Int. Gear	2300	80.2	116	210
53.8	Int. Gear	2300	101.0	146	210
67.2	Int. Gear	2300	124.8	181	210
86.1	Int. Gear	1800	163.0	236	210
107.6	Int. Gear	1800	201.3	291	210
134.5	Int. Gear	1500	249.2	361	210

Pump Data – Internal Gear Pumps

Table A1.13 | Internal gear pumps (Pressure Range - 320 bar)

Motor kW	Pump type	Speed (rpm)	Displacement (cc/rev)	Flow rate (lpm)	Pressure (Cont) (bar)
4.1	Int. Gear	5000	5.1	7.4	320
4.9	Int. Gear	5000	6.3	9.1	320
6.6	Int. Gear	5000	8.0	11.5	320
8.2	Int. Gear	4300	10.0	14.5	320
9.8	Int. Gear	4300	12.6	18.3	320
13.1	Int. Gear	4300	15.6	22.6	320
16.4	Int. Gear	3600	20.4	29.5	320
20.5	Int. Gear	3600	25.1	36.4	320
26.2	Int. Gear	3600	32.4	46.8	320
32.8	Int. Gear	3000	39.3	56.9	320
41.0	Int. Gear	3000	50.6	73.2	320
51.6	Int. Gear	3000	63.7	92.1	320
65.6	Int. Gear	2300	80.2	116	320
82.0	Int. Gear	2300	101.0	146	320
102.5	Int. Gear	2300	124.8	181	320
131.2	Int. Gear	1800	163.0	236	320
163.9	Int. Gear	1800	201.3	291	320
205.0	Int. Gear	1500	249.2	361	320

Pump Data – Gerotor Pumps

Table A1.14 | Gerotor pumps

Motor kW	Pump type	Max. Speed (rpm)	Displacement (cc/rev)	Flow rate (lpm) [Per 1000 rpm]	Pressure (Cont) (bar)
5.4	Gerotor	5000	3.57	3.6	138
9.1	Gerotor	5000	6.09	6.1	138
11.1	Gerotor	5000	7.37	7.4	138
14.2	Gerotor	5000	9.5	9.5	138
12.9	Gerotor	5000	11.47	11.5	103.5

Pump Data – Vane Pumps

Table A1.15 | Vane pumps

Motor kW	Pump type	Speed (rpm)	Displacement (cc/rev)	Flow rate (lpm)	Pressure, operating (bar)	Pressure, Rated (bar)
2	Vane	1500	18	22	35	210
2	Vane	1800	18	27	35	210
3	Vane	1500	27	35	35	210
3	Vane	1800	27	42	35	210
4	Vane	1500	18	22	70	210
4	Vane	1800	18	27	70	210
4	Vane	1500	36	48	35	210
4	Vane	1800	36	58	35	210
4	Vane	1500	40	52	35	175
5	Vane	1800	40	62	35	175
5	Vane	1500	45	58	35	175
6	Vane	1500	18	22	100	210
6	Vane	1500	27	35	70	210
6	Vane	1800	45	70	35	175
6	Vane	1500	55	72	35	175
7	Vane	1500	18	22	140	210
7	Vane	1800	18	27	100	210
7	Vane	1800	27	42	70	210
7	Vane	1500	36	48	70	210
7	Vane	1800	55	86	35	175
7	Vane	1500	67	90	35	175
8	Vane	1500	27	35	100	210
8	Vane	1500	40	52	70	175
8	Vane	1800	67	108	35	175
9	Vane	1500	18	22	175	210
9	Vane	1800	18	27	140	210
9	Vane	1800	36	58	70	210
9	Vane	1500	45	58	70	175
10	Vane	1800	27	42	100	210
10	Vane	1800	40	62	70	175
11	Vane	1500	18	22	210	210
11	Vane	1800	18	27	175	210
11	Vane	1500	27	35	140	210
11	Vane	1500	36	48	100	210
11	Vane	1800	45	70	70	175
11	Vane	1500	55	72	70	175

Pump Data – Vane Pumps, Contd...

Table A1.16 | Vane pumps

Motor kW	Pump type	Speed (rpm)	Displacement (cc/rev)	Flow rate (lpm)	Pressure, operating (bar)	Pressure, Rated (bar)
12	Vane	1500	40	52	100	175
13	Vane	1800	18	27	210	210
13	Vane	1800	27	42	140	210
13	Vane	1800	36	58	100	210
14	Vane	1500	27	35	175	210
14	Vane	1500	45	58	100	175
14	Vane	1800	55	86	70	175
14	Vane	1500	67	90	70	175
15	Vane	1500	36	48	140	210
15	Vane	1800	40	62	100	175
16	Vane	1500	27	35	210	210
16	Vane	1800	27	42	175	210
16	Vane	1500	40	52	140	175
27	Vane	1200	126	121	103	
31	Vane	1200	162	159	103	
37	Vane	1800	126	189	103	
45	Vane	1800	162	265	103	

Table A1.17 | Vane pumps

Motor kW	Pump type	Max. Speed (rpm)	Displacement (cc/rev)	Flow rate (lpm) @1500 rpm/7bar	Pressure (Cont) (bar)
5.6	Vane	4800	3.3	4.7	175
9.6	Vane	4800	5.5	7.9	175
10.5	Vane	4500	6.5	9.4	175
14.2	Vane	4000	9.8	14.2	175
16.0	Vane	3400	13.1	18.9	175
19.0	Vane	3200	16.4	23.6	175
18.2	Vane	3000	19.5	28.4	150
18.5	Vane	2800	22.8	33.1	140

Pump Data – Piston Pumps

Table A1.18 | Axial piston pumps

Motor kW	Pump type	Speed (rpm)	Displacement (cc/rev)	Flow rate (lpm)	Pressure, operating (bar)	Pressure, Rated (bar)
3	Piston	1500	20	28	35	210
3	Piston	1800	20	34	35	210
4	Piston	1500	20	28	70	210
4	Piston	1500	32	48	35	140
5	Piston	1800	32	58	35	140
5	Piston	1500	40	58	35	210
5	Piston	1500	45	63	35	186
5	Piston	1800	45	76	35	186
6	Piston	1800	40	67	35	210
7	Piston	1500	20	28	70	210
7	Piston	1500	32	48	70	140
7	Piston	1500	57	82	35	250
8	Piston	1800	20	34	100	210
8	Piston	1800	32	58	70	140
9	Piston	1500	20	28	140	210
9	Piston	1500	40	58	70	210
9	Piston	1500	45	63	70	186
9	Piston	1500	63	92	35	210
9	Piston	1500	74	110	35	250
10	Piston	1500	32	48	100	140
10	Piston	1800	57	98	35	250
11	Piston	1500	20	28	175	210
11	Piston	1800	20	34	140	210
11	Piston	1800	40	67	70	210
11	Piston	1800	45	76	70	186
11	Piston	1800	63	110	35	210
11	Piston	1500	81	117	35	210
12	Piston	1800	32	58	100	140
12	Piston	1500	57	82	70	250
12	Piston	1800	74	132	35	250
13	Piston	1800	20	34	175	210
13	Piston	1500	40	58	100	210
13	Piston	1800	81	140	35	210
13	Piston	1500	98	150	35	250
14	Piston	1500	20	28	210	210
14	Piston	1500	32	48	140	140

Pump Data – Piston Pumps, Contd...

Table A1.19 | Axial piston pumps

Motor kW	Pump type	Max. Speed (rpm)	Displacement (cc/rev)	Flow rate (lpm) @ rated rpm	Pressure, operating (bar)	Pressure, Rated (bar)
14	Piston	1500	45	63	100	186
14	Piston	1500	63	92	70	210
14	Piston	1500	106	152	35	210
15	Piston	1800	40	67	100	210
15	Piston	1800	57	98	70	250
15	Piston	1500	74	110	70	250
16	Piston	1500	40	58	140	210
16	Piston	1800	98	170	35	250

Table A1.20 | Axial piston pumps, variable

Motor kW	Pump type	Speed (rpm)	Displacement (cc/rev)	Flow rate (lpm)	Pressure (Cont) (bar)
2.2	Piston, variable	1420	25.0	33.7	35
3.0	Piston, variable	1420	25.0	33.7	50
4.0	Piston, variable	1440	25.0	34.1	65
5.5	Piston, variable	1455	25.0	34.4	90
5.5	Piston, variable	1455	38.0	52.5	55
7.5	Piston, variable	1455	25.0	34.4	120
7.5	Piston, variable	1455	38.0	52.5	75

Pump Data – Piston Pumps, Contd...

Table A1.21 | Bent-axis piston pumps

Motor kW	Pump type	Speed (rpm)	Displacement (cc/rev)	Flow rate (lpm)	Pressure, operating (bar)	Pressure, Rated (bar)
50	Bent-axis	2500	40.9	92.0	350	400
63	Bent-axis	2400	50.1	120.2	350	400
65	Bent-axis	2200	63	138.6	350	400
68	Bent-axis	2000	71.6	143.2	315	350
74	Bent-axis	2000	78.3	156.6	315	350
83	Bent-axis	2000	79.1	158.2	350	400
100	Bent-axis	2000	110	220.0	300	350

13 | References

1. Article on (1) 'About Hydraulic Pumps' and (2) on 'About Hydraulic Reservoirs', GlobalSpec Inc., Jordan Rd, Troy, NY, USA.

2. Articles on 'Hydraulic reservoirs' and 'Heat exchangers' Penton Media, Inc. & Hydraulics & Pneumatics magazine

3. Catalogues on 'Reservoirs', 'In tank-coolers (water-cooled)', 'Suction strainers', 'Tank magnets', VESCOR, LDI Industries, Manitowoc, WI.

4. Cooling Product Catalogue, E 57.000.1 / 08.16, HYDAC International, Industriegebiet, 66280 Sulzbach/Saar, Germany

5. Document on 'Hydraulic Filtration Product Guide', Catalog No. F112100 ENG (2/12), Donaldson Company, Inc. PO Box 1299 Minneapolis, MN 55440-1299

6. Document on 'Hydraulic Power Unit HV350AS', CCI, USA

7. Document on Anderson Greenwood Type 81P SOPRV Installation and Maintenance Instructions of Spring Operated Safety Relief Valves (SOPRV), www.tycovalves.com

8. Document on Crosby® Pressure Relief Valve Engineering Handbook, Technical Document No. TP-V300, Anderson Greenwood, Crosby Valve Inc.

9. Document on: 'Hydraulic Power Pack SA 4', SA4_7100_1en_02/2016, AGRO HYTOS, www.argo-hytos.com

10. Document on: 'Overview of Reservoir Accessories' HYDAC, Germany

11. Owner's Manual - Hydraulic Reservoirs, Parker Hannifin Corporation, Mississippi, USA.

12. Vickers literature #510, 'Noise Control in Hydraulic Systems, for design guidelines'.

Fluid Power Educational Series Books

1. Pneumatic Systems and Circuits -Basic Level (In the SI Units)
2. Industrial Pneumatics -Basic Level (In the English Units)
3. Pneumatic Systems and Circuits -Advanced Level
4. Electro-Pneumatics and Automation
5. Design of Pneumatic Systems (In the SI Units)
6. Design Concepts in Pneumatic Systems (In the English Units)
7. Maintenance, Troubleshooting, and Safety in Pneumatic Systems
8. Industrial Hydraulic Systems and Circuits -Basic Level (In the SI Units)
9. Industrial Hydraulics -Basic Level (In the English Units)
10. Hydraulic Fluids
11. Hydraulic Filters: Construction, Installation Locations, and Specifications
12. Hydraulic Power Packs (In the SI Units)
13. Power Packs in Hydraulic Systems (In the English Units)
14. Hydraulic Cylinders (In the SI Units)
15. Hydraulic Linear Actuators (In the English Units)
16. Hydraulic Motors (In the SI Units)
17. Hydraulic Rotary Actuators (In the English Units)
18. Hydraulic Accumulators and Circuits (In the SI Units)
19. Accumulators in Hydraulic Systems (In the English Units)
20. Hydraulic Pipes, Tubes, and Hoses (In the SI Units)
21. Pipes, Tubes, and Hoses in Hydraulic Systems (In the English Units)
22. Design of Industrial Hydraulic Systems (In the SI Units)
23. Design Concepts in Industrial Hydraulic Systems (In the English Units)
24. Maintenance, Troubleshooting, and Safety in Hydraulic Systems
25. Hydrostatic Transmissions (HSTs) (In the SI Units)
26. Concepts of Hydrostatic Transmissions (In the English Units)
27. Load Sensing Hydraulic Systems (In the SI Units)
28. Concepts of Load Sensing Hydraulic Systems (In the English Units)
29. Electro-hydraulic Proportional Valves
30. Electro-hydraulic Servo Valves
31. Cartridge Valves
32. Electro-hydraulic Systems and Relay Circuits

For more details, please visit: **htpps://jojibooks.com**

About the Author

Joji Parambath is a trainer in the field of Pneumatics, Hydraulics, and PLC, for over 25 years. During his career, he has trained numerous professionals from the industries as well as faculty members and students of engineering institutions.

At present, he is the key trainer at Fluidsys Training Centre, Bangalore, India, (https://fluidsys.org) which is providing training in the field of Pneumatics and Hydraulics. He has already written two books on Pneumatics and Hydraulics. The publication of the present series of 32 books is intended to restructure and update the existing books.

The author wishes to thank all trainees for their lively interaction and many useful suggestions during the training programmes that prompted the author to write the present series of books. You may send your feedback to joji.p@hotmail.com

10th June 2020

nod-product-compliance